Recent Advancements in Objective Geology

Recent Advancements in Objective Geology

Dr. Brahmanand Sharma

Kalpana Mishra

Aditya Singh Baghel

KALPAZ

ISBN: 978-93-5324-937-3

First Published, 2022

Published by

Kalpaz Publications
C-30, Satyawati Nagar,
Delhi – 110052
Ph.: 011-47034999, 9811692060
www.kalpazpublications.com
E-mail: kalpaz@hotmail.com

Printed at: Gyan Press, Delhi.

Recent Advancements
in
Objective Geology

Dedicated to

Our Respected Teacher

Dr. Rabindra Nath Tiwari

Professor of Geology

Government Model Science College

Rewa (M.P.)-486001

CONTENTS

TERMS

- The publisher/seller/printer for any part of this book does not warrant that the information's given in this book is complete in the final sense.
- If the readers on any part or question of this book, then they should assess it on the basic of their source and if any discrepancy is found in this book, then tell the discrepancy at the Authors / Publisher address so that this book will be improved in the upcoming edition.
- Full care has been taken by the publisher in publishing this book, yet the publisher / seller / authors / will not be responsible for any loss or damage caused to the student due to any error.
- Jurisdiction for any dispute will be Rewa Madhya Pradesh.

PREFACE

Earth Science is one of the most demanding subjects in various aspects for students and professionals. Geology deals with the study of earth in a holistic manner. It is also very useful subject with social and economical point of view. Various Universities and Colleges offer Bachelor, Master degree as well as Ph. D degree in the Geology. Even at Higher Secondary level, some of the states are planning to introduce this subject. The main branches of Geology are- Geodynamics, Geomorphology, Mineralogy, Petrology, Structure Geology, Ore Geology, Palaeontology, Stratigraphy, Hydrogeology, Remote Sensing, Mining and Engineering etc.

The present book is devoted to Recent Advancements in Objective Geology which covers all branches of Geology with more than 1700 objective questions and their answers. The book may be very useful for UPSC Exam (GSI), NET, GATE, ONGC Exam, States PSCs, States TET, Researchers, P.G. Entrance exam etc.

To publish such an important book for students, academicians, researchers and various organisations was a tough task. We are very grateful to almighty God to give us courage to bring out this book. We are extremely grateful to our respected Sir Prof. Rabindra Nath Tiwari, who inspired and motivated us to publish this book.

We are thankful to Prof. R. P. Tiwari, Honourable Vice Chancellor, Central University, Punjab; Dr. P. K. Shrivastav, Additional Director, Higher Education Rewa Division; Shri Niteesh Kumar Dutta, Ex. Director General, GSI; Dr. D. P. Dubey, Rewa; Prof. G.P. Pandey, Rewa; Dr. D. K. Singh, Rewa; Dr. Sanjay Tignath, Jabalpur; Dr. Ramesh Shrivastav, Satna; Dr. R. K. Shrivastav, MVM, Bhopal; Shri S. K. Sarkar, Ex Director, GSI; Shri M. K. Soni, Ex Deputy Director General (GSI); Dr. Arun Tripathi, Rewa; Dr. B. K. Mishra, A.K.S. University, Satna; Mr. Varun Mishra, NMDC, Panna; Mr. Vijay Soni, Rewa; Mr. Bhupendra Kumar Chandrakar, Dy. Director (Mining Administration), Chhattisgarh; Mr. Suman Das, Director, Geosumukh Inframines Private Limited; Mr. Sujan Singh Lodhi, Mauganj; Dr. Abha Dubey, Rewa; Dr. Parul Dubey, Indore; Dr. Umesh

Mishra, Rewa; Dr. Amit Singh, Rewa; Mr. Pushpendra Tiwari, Rewa; Dr. Vidyakant Tiwari, Mining Inspector; Dr. Ajay Mishra, Mining Inspector; Dr. Amit Mishra, Rewa; Dr. Praveen Tiwari, Rewa; Dr. Rashmi Dwivedi, Rewa; Dr. Shrutika Singh, Rewa; Sana Siddique, Rewa; Heeral Mishra, Rewa; Divya Singh, Rewa for their encouragement and suggestions.

We have tried our best to publish excellent quality book; however some errors may be there in the book. The suggestions and recommendations by the readers are always welcome in this direction.

CHAPTER - 1

GEODYNAMICS

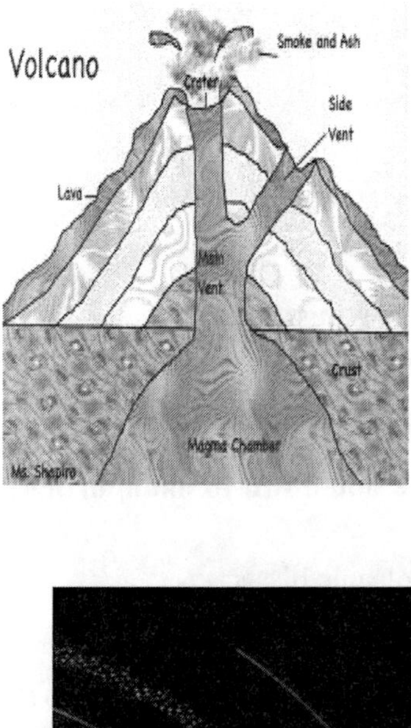

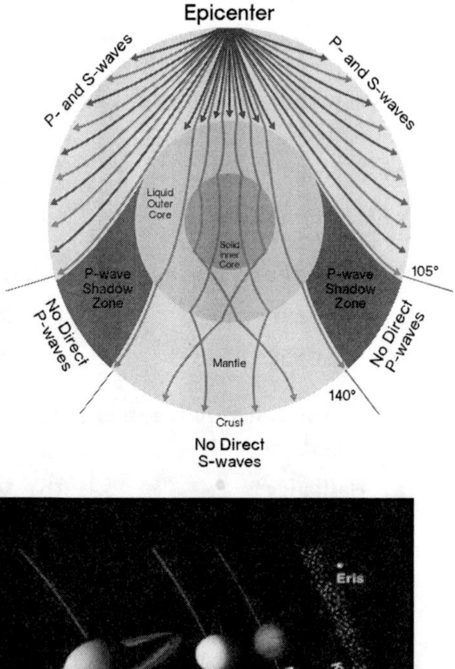

Geodynamics

1. **Which name is associated with "Principle of Uniformitarianism"?**
   (a) William Whewell (b) James Hutton
   (c) John Playfair (d) Charles Lyell

2. **Highly viscous, mechanical week and ductile upper part of the mantle of the Earth that lies at depths between 80 – 200 km is composed chiefly of:**
   (a) Dunite (b) Spinel peridotite
   (c) Pyroxenite (d) Peridotite

3. **With which of the following processes is sea-floor spreading connected?**
   (a) Igneous Activity (b) Metamorphism
   (c) Earth movements (d) Denudation

4. **Which of the following is the fastest moving plate?**
   (a) Eurasian plate (b) Indian plate
   (c) Antarctica plate (d) Pacific plate

5. **The finest volcanic fragments from 2 mm down to nanoparticles are called:**
   (a) Ballistics (b) Volcanic blasts
   (c) Lahars (d) Tephra

6. **Steep sided volcanoes which are mixtures of lava and pyroclastic deposit are called:**
   (a) Shield volcanoes (b) Stratovolcanoes
   (c) Calderas (d) Somma

7. **What is the total number of volcanoes that erupted in the last 10,000 years?**
   (a) 866 (b) 1,551
   (c) 347 (d) 2,208

8. **Heat flow of the Earth is commonly measured in Heat Flow Units (HFU), for which 1 HFU is equal to:**
   (a) 39.84 mW/m² (b) 40.84 mW/m²
   (c) 41.83 mW/m² (d) 42.8 mW/m²

9. **Matuyama reversal occurred between:**
   (a) 0.1 – 0.69 Ma (b) 0.69 – 2.43 Ma
   (c) 2.43 – 3.32 Ma (d) 3.32 – 3.5 Ma

10. **With which of the following events is Gauss normal epoch coinciding?**
    (a) Mammoth (b) Gilsa
    (c) Cochiti (d) Olduvi

11. **Who proposed the "Oscillation Theory" to explain orogenesis?**
    (a) Bird (1970) (b) Haarmann (1930)
    (c) Gignoux (1937) (d) Wegmann (1932)

12. **In a continental shelf, grain size is sorted by wave action, down to a depth of about:**
    (a) 20 m (b) 30 m
    (c) 40 m (d) 50 m

13. **What is the mean density of the Earth?**
    (a) 5.967 g/cm³ (b) 5.639 g/cm³
    (c) 5.239 g/cm³ (d) 5.517 g/cm³

14. **Which of the following submarine mountain chains extends around the Earth as mid oceanic ridges that rise to heights of 3000-4000 m above the ocean plains and is located 200 km SW of Bombay coast?**
    (a) Mid Indian Ridge (b) Antarctica Ridge
    (c) Carlsberg Ridge (d) Atlantic Ridge

15. **Which of the most convincing geological evidence for continental drift?**
    (a) Paramagnetic (b) Paleontological
    (c) Radiometric ages
    (d) Permo-carboniferous glaciations of Gondwanaland

16. Magnitude of a strong earthquake ranges between:

(a) 7.0- 7.3 (b) 6.2 – 6.9

(c) 5.5 – 6.1 (d) 4.9 – 5.4

17. What is the velocity of the 'P' waves at the upper and lower mantle boundary?

(a) 8.1 Km/s (b) 8.97 Km/s

(c) 10.70 Km/s (d) 11.42 Km/s

18. The northern prow of the advancing Indian shield during late Cretaceous and tertiary orogenic cycles extends beneath which of the following?

(a) Deccan plateau (b) Indo-Gangetic plain

(c) Tibet-Altyntagh (d) Gobi Desert

19. Which of the following is different from others in composition?

(a) Mars (b) Mercury

(c) Saturn (d) Venus

20. The layer just below the lithosphere is known as ....

(a) Low velocity zone

(b) Gutenberg discontinuity

(c) The zone of deep-seated Earthquakes

(d) Inversion zone

21. New ocean floor is created at......

(a) Deep Sea Trenches (b) Subduction Zones

(c) Mid Oceanic Ridges (d) Island Arcs

22. Iceland can be cited as an example of –

(a) Oceanic Trenches (b) Subduction Zones

(c) Oceanic Island (d) Mid Ocean Ridge

23. What is the name of permanent Indian Station in East Antarctica?

(a) Dakshin Gangotri (b) Maitri

(c) Bharti (d) All the above

24. Which of the following type of seismic waves arrives first at the seismic station from an earthquake hypocenter?

(a) P-waves (b) S-waves

(c) Rayleigh waves (d) Love waves

25. Which of the following represents the correct order of density of crust – mantle – core and bulk earth respectively (in gm/cm^3)

(a) 2.8 – 4.5 - 11.0 – 5.5

(b) 2.8 - 5.5 – 11.0 – 4.5

(c) 2.8 - 11.0 – 5.5 - 4.5

(d) 2.8 – 4.5 – 5.5 – 11.0

26. In which one of the following mountains ranges volcanic rocks are least common?

(a) Andes (b) Cascades

(c) Himalayas (d) Sierra Nevada's

27. Shadow zones for direct P- and S - waves lies between-

(a) 102° to 142° for both P- and S- waves

(b) 102° to 180° for P- wave and 102° to 142° for S- waves

(c) 102° to 180° for both P- and S- waves

(d) 102° to 142° for P-wave and 102° to 180° for S-waves

28. Polymetallic nodules on the ocean floor contain significant amount of:

(a) Cu-Ni-Co (b) Pb - Zn - To

(c) Hg - Mo- Pt (d) U -Th - Nb

29. The average radius of the Earth is –

(a) 5420 Km (b) 6370 Km

(c) 9880 Km (d) 3340 Km

30. Oldham Discontinuity marks the boundary between –

(a) Crust and Upper Mantle

(b) Mantle and Outer Core

(c) Outer Core and Inner Core

(d) Upper Mantle and Inner Mantle

31. P-waves shadow zones recorded by seismologists indicate the boundary between the

(a) Lower and upper crust

(b) Inner core and outer core

(c) Lower mantle and outer core

(d) Transition zone and lower mantle

32. The mafic minerals contain more of

(a) Al and Si (b) Mg and Fe

(c) Ca and Cu (d) Mn and Al

33. The epic entre of a deep focus earthquake is located

(a) On the earth's surface (b) Below a depth of 70 Km

(c) Between 70-300 Km (d) Below 300 Km

34. Which one of the following is an example of an isostatic movement?

(a) Stream down cutting following a drop in sea level

(b) Arching of strata at the center of a dime

(c) Numerous aftershocks associated with deep focus earthquake

(d) Uplift of areas recently covered by thick, continental ice sheets

35. In convergent plate tectonic setting, thickest continental lithosphere is associated with

(a) Ocean- Ocean convergent setting

(b) Ocean-continent convergent setting

(c) Continent-Continent collision setting

(d) Ocean-ocean divergent setting

36. Which of the following subdivision of the Himalaya is constituted of crystalline metamorphic rocks only

(a) Lesser Himalaya (b) Tethys Himalaya

(c) Outer Himalaya (d) Greater Himalaya

37. Which of the following shows the correct order of lithotectonic subdivision of the Himalaya from South to North

(a) Lesser Himalaya- Outer Himalaya- Tethys Himalaya- Greater Himalaya

(b) Outer Himalaya- Lesser Himalaya- Greater Himalaya- Tethys Himalaya

(c) Lesser Himalaya- Outer Himalaya- Greater Himalaya- Greater Himalaya

(d) Outer Himalaya- Outer Himalaya- Tethys Himalaya- Greater Himalaya

38. The main Boundary thrust separates

(a) Greater Himalaya from Tethys Himalaya

(b) Lesser Himalaya from Greater Himalaya

(c) Outer Himalaya from Lesser Himalaya

(d) Outer Himalaya from Gangetic Himalaya

39. Upper Tertiary rocks dominate in:

(a) Tethys Himalaya (b) Outer Himalaya

(c) Greater Himalaya (d) Lesser Himalaya

40. The hypothesis the life on Earth plays an important role in regulating the planetary climate is known as:

(a) Gaia hypothesis (b) Uniformitarianism

(c) Sustainability hypothesis (d) Harry Hess hypothesis

41. Oceanic waters up to 100 m are dominated by:

(a) Cibicides like forms (b) Agglutinates forms

(c) Spinose planktic forms (d) Porellaneous forms

42. One of the following, who is associated with the Law of Superposition?

(a) Nicholas Steno (b) Fred Hoyle

(c) Lehman (d) James Hutton

43. Life on the earth depend on the following interconnected factored:

1. High quality energy from the sun
2. Cycling of matter
3. Gravity
4. Six elements
5. Water

Option:

(a) (1), (2), (5) (b) (1), (2), (3)

(c) (2), (3), (4) (d) (4), (2), (1)

44. The earth would be the same as other planets in the solar system without:

(a) Volcanoes and plate tectonics
(b) Water and rocks
(c) River and plants
(d) Plants and animals

45. Thickness of the earth's upper mantle is up to:

(a) 100 km (b) 420 km
(c) 670 km (d) >100 km

46. Adiabatic decompression melting involves melting at:

(a) Low T and high P (b) High T and low P
(c) Constant T but increasing P (d) Constant T but decreasing P

47. Nearly 50% of the Earth's mass consist of:

(a) Upper mantle (b) Transition zone
(c) Lower mantle (d) Outer core

48. Role of H_2O in melting is:

(a) Increase P at a given T (b) Increase T at a given P
(c) Decrease P at a given T (d) Decrease T at a given P

49. Perovskite occurs in:

(a) Upper mantle of the earth (b) Transition zone of the earth
(c) Lower mantle of the earth (d) Inner core of the earth

50. "D" layer is identified at the boundary of:

(a) Lower mantle and another core
(b) Crust and Mantle
(c) Upper and lower mantle
(d) Outer and inner core

51. Near earth's surface condition:

(a) Olivine is more stable than amphibole
(b) Quartz is less stable than clay minerals
(c) Pyroxene is less stable than dolomite
(d) Muscovite is more stable than quartz

52. Daughter nucleus has atomic number 2 less and mass number 4 less than parent due to:

(a) Alpha decay (b) Beta decay
(c) Electron capture decay (d) Gamma decay

53. The average composition of the continental crust is similar to that of:

(a) Granite (b) Granodiorite
(c) Diorite (d) Basalt

54. Crustal shortening takes place due to:

(a) Normal faulting (b) Folding and thrusting
(c) Strike – slip faulting (d) Intrusion

55. The discontinuity between crust and Mantle is called

(a) Mohorovicic (b) Lehmann
(c) Gutenberg (d) Karlsbad

56. Which of the following is not extruded by a volcano

(a) Ash (b) Magma
(c) Bomb (d) Lava

57. The doctrine of Uniformitarianism was given by

(a) Nicholas Steno (b) James Hutton
(c) William Smith (d) Charles

58. Which of the following marks the boundary between the continental crust and Oceanic crust?

(a) Continental slope (b) Continental rise
(c) Continental shelf (d) Abyssal plain

59. Which of the following is responsible for movement of water through Oceanic currents

(a) Salinity
(b) Salinity, temperature and density
(c) Viscosity and debris content
(d) Presence of volcanoes

60. Interstadials during quaternary period are

(a) Long warm events between the generally colder periods

(b) Short warm events between the generally colder periods
(c) Short cold events between the generally warmer periods
(d) Long warm events only

61. What is measured with Richter scale?

(a) Earthquake magnitude (b) Earthquake focus
(c) Earthquake intensity (d) Earthquake epicentre

62. Which of the following is a part of the Atmosphere?

(a) Biosphere (b) Asthenosphere
(c) Stratosphere (d) Lithosphere

63. Another name for Seismic sea waves is

(a) Typhoon (b) Tsunami
(c) Current (d) Cyclone

64. The average depth of the Mohorovicic Discontinuity is

(a) 5 Km (b) 15 Km
(c) 20 Km (d) 35 Km

65. Most elements in the earth's surface exist

(a) In oxide state (b) In reduced state
(c) In metal state (d) In vapour state

66. Pangaea is the name of a farmer content that included

(a) All the contents
(b) Only Europe and Asia
(c) Only South America and Africa
(d) Only North and South America

67. The core of the earth age thought to be composed of which of the following

(a) Basalt and grenade (b) Compressed gases
(c) Uranium and lead (d) Iron and nickel

68. Which of the following distance in kilometers is closest to the diameter of the earth?

(a) 750 kilometers (b) 2750 kilometers
(c) 12750 kilometers (d) 127500 kilometers

69. The Mohorovicic Discontinue, commonly called the Moho, is the boundary between the?

(a) The mantle and the crust
(b) Asthenosphere and the mantle
(c) Mantle and the outer core
(d) Outer core and the inner core

70. The oxygen in the earth's atmosphere was probably produced by which of the following?

(a) Volcanoes
(b) Photosynthetic organisms
(c) Radioactive minerals
(d) Meteors

71. According to the protoplanet hypothesis, the solar system begins as which of the following?

(a) A start
(b) A Vacuum
(c) A huge cloud of dust and gas
(d) A group of comets

72. Continents have three major features. The oldest area of each continent or the:

(a) Folded mountains
(b) Stable platforms
(c) Mountain course
(d) Shields

73. Which of the following names is given to the supercontinents thought to have existed in the Cambrian period?

(a) Pangaea
(b) Afrasia
(c) Gondwanaland
(d) Europia

74. In the nebular hypothesis, each planet is formed from which of the following?

(a) A comet
(b) A number of moons
(c) A passing star
(d) A ring of gas

75. The earth's average density is

(a) 2.82 grams per cubic centimeter
(b) 4.75 grams per cubic centimeter
(c) 5.52 grams per cubic centimeter
(d) 10.5 grams per cubic centimeter

76. The tectonic plate on which India and the adjacent portion of Asia ride are colliding at a rate of:

(a) 1-10 Millimeters per year

(b) 1-10 Centimeters per year

(c) 1-10 Meters per year

(d) 1-10 Decimeters per year

77. What is the name given to a region where one plate of the earth's crust is sliding under another?

(a) Subduction zone
(b) Convection zone
(c) Collision zone
(d) Transform fault

78. We know that the earth's outer core is liquid because:

(a) P waves pass through it

(b) S waves pass through it

(c) P waves cannot pass through it

(d) S waves cannot pass through it

79. Which of the following is the name of the partly melted rock layer on which the tectonic plates move?

(a) Lithosphere
(b) Asthenosphere
(c) Hydrosphere
(d) Outer core

80. A volcano which is composed of lava flows and pyroclastic material and which is steep-sided and very tall is known as:

(a) Syncline
(b) Composite cone
(c) Anticline
(d) None of these

81. Which of the following energy sources does NOT originally come from the sun?

(a) Wind energy
(b) Ocean thermal energy
(c) Geothermal energy
(d) Hydroelectric energy

82. What percentage of the earth's atmosphere does O_2 comprise?

(a) 75%
(b) 50%
(c) 21%
(d) 32%

83. What gas makes up the majority of our atmosphere?

(a) Oxygen (b) Carbon dioxide

(c) Ozone (d) Nitrogen

84. In geological studies, all of the following are TRUE of P waves except that they:

(a) Are body waves (b) Travel like sound waves

(c) Can pass through liquids (d) Are slower than S waves

85. What percentage of the earth's water is fresh water?

(a) 50% (b) 22%

(c) 7% (d) 3%

86. Which of the following is TRUE of an earthquake's epicentre and focus?

(a) They are the same

(b) The focus is above the epicenter

(c) The epicentre is at the earth's surface

(d) The focus is at the earth's inner core

87. Most earthquake originate at depths between:

(a) The surface and 100 kilometers

(b) 100 kilometres and 500 kilometers

(c) 500 kilometres and 700 kilometers

(d) 700 kilometres and the outer core

88. Waves generated by underwater earthquake are called:

(a) Tsunamis (b) A bore

(c) A tidal wave (d) Microwave

89. The principal volcanic gas (in terms of quantity) is:

(a) Carbon dioxide (b) Nitrogen dioxide

(c) Sulfur dioxide (d) Steam (H_2O)

90. Which one of the following observations supports The Big Bang Theory?

(a) Recession of galaxies

(b) Expansion of galaxies

(c) Recession and Expansion of galaxies
(d) Stagnation of galaxies

91. Which one of the following radioactive isotope methods is not suitable for dating the Precambrian rocks?

(a) 14C (b) Rb/Sr method
(c) Sm/Nd method (d) U-Pb method

92. Which one of the following represents composition of the testacy-Iron meteorites?

(a) Nickel, Iron and Sulphates (b) Nickel, Iron and Phosphates
(c) Nickel, Iron and Silicates (d) Nickel, Iron and Carbonates

93. Which one of the following pairs makes the bulk composition of the sun?

(a) Hydrogen and hydrogen (b) Hydrogen and helium
(c) Hydrogen and argon (d) Helium and argon

94. Which one of the following is related to the development of rugged topography of mid oceanic ridges?

(a) Fast spreading rate (b) Intermediate spreading rate
(c) Slow spreading rate (d) Uncertain spreading rate

95. Which one of the following is the correct sequence of events in the Wilsonian Cycle?

(a) Breakup of continents - Seafloor spreading - Subduction collision
(b) Sea floor spreading - Breakup of continents - Subduction collision
(c) Breakup of continents - Subduction collision - Seafloor spreading
(d) Subduction collision - Seafloor spreading - Breakup of continents

96. Which one of the following is a plate boundary as well as a strike-slip fault?

(a) Himalayan Frontal fault (b) Great Boundary fault
(c) Great Glen fault (d) San Andreas fault

97. Who proposed that the convention currents in asthenosphere cause the lithospheric plate movement?

(a) Arthur Holmes (b) Tuzo Wilson
(c) M.P. Billings (d) D.N. Wadia

98. The average of radius of the Earth is –

(a) 5420 Km (b) 6370 Km

(c) 9880 Km (d) 3340 Km

99. Oldham Discontinuity marks the boundary between –

(a) Crust and Upper Mantle

(b) Mantle and Outer Core

(c) Outer Core and Inner Core

(d) Upper Mantle and Inner Mantle

100. The average thickness of the Earth's crust in –

(a) 193 km (b) 1233 km

(c) 33 km (d) 1238 km

101. The 'Seismic-shadow-zone-angle' that exists from the epicentre is –

(a) 48° to 92° (b) 105° to 143°

(c) 150° to 165° (d) 165° to 178°

102. Seismic P-Waves can travel through –

(a) Crust only (b) Mantle only

(c) Crust and Mantle only (d) Crust, Mantle and Core

103. Oldham Discontinuity is identified on the basis of –

(a) Sudden increase in the velocity of P-waves

(b) Sudden decrease in the velocity of P and S waves

(c) Sudden increase in the velocity S-waves

(d) Sudden increase in the velocity P & S waves

104. Primary seismic wave propagation is medium causes -

(a) Vibration of particles in the wave propagation direction

(b) Vibration of particles transverse to the wave propagation direction

(c) Particles do not vibrate

(d) Particle may vibrate in any direction

105. Oceanic crust has dominance of which elements

(a) Sodium and Potassium (b) Silicon and Magnesium

(c) Calcium and Nickel (d) Chromium and Tungsten

106. Continental crust has dominance of which elements -

(a) Iron and Chromium (b) Silicone and Aluminium

(c) Calcium and Magnesium (d) Nickel and Sodium

107. The super Continent”Rodinia” existed during which time -

(a) ArcheanTime (b) Palaeo Proterozoic Time

(c) Meso Proterozoic Time (d) Neo Proterozoic Time

108. The most abundant element in the crystal rocks, is -

(a) Silicon (b) Oxygen

(c) Iron (d) Calcium

109. The core of the Earth in made up of -

(a) Nickel and Chromium (b) Nickel and Iron

(c) Iron and Chromium (d) Iron and Calcium

110. Total number of Rare Earth Elements (REE) is -

(a) 15 (b) 20

(c) 17 (d) 24

111. What type of wave generates at the time of an Earthquake -

(a) Electromagnetic waves (b) Elastic waves

(c) Gamma waves (d) Gravitation waves

112. What is a “Seismic Gap”?

(a) The time difference between the successive high magnitude earthquake at one place

(b) The segment of an active fault, where significant earthquake had not occurred after considerable time

(c) Opening of large voids in rocks due to earthquake

(d) En-echelon open crack delivered due to earthquake

113. Which one of the following is a plate boundary associated commonly with the paired metamorphic belt?

(a) Mid oceanic ridge (b) Transform fault

(c) Subduction zone (d) Continental rift

114. Who has proposed the Principle of ‘uniformitarism’?

(a) James Hutton (b) Nicolaus Steno

(c) Charles Lyell (d) William Smith

115. The surface waves are

(a) P-waves
(b) S-waves
(c) Both P and S waves
(d) L-waves

116. Conrad discontinuity separates boundary between

(a) Lower crust and upper mantle
(b) Upper mantle and lower mantle
(c) Outer core and inner core
(d) Upper crust and lower crust

117. The new oceanic surface is created at .... plate margin.

(a) Constructive
(b) Distractive
(c) Conservative
(d) Convergent

118. Match the name of scientist (Group-I) and their famous theirs (Group-II)

Group-1	**Group-2**
P. Alfred	1. Isostasy
Q. Airy	2. Sea floor spreading
R. Herry Hess	3. Theory of continental drift
S. Pratt	4. Big Bang theory

(a) P-1, Q-2, R-3, S-4
(b) P-3, Q-1, R-2, S-4
(c) P-4, Q-1, R24, S-3
(d) P-1, Q-2, R-4, S-3

119. Which of the following is supercontinent?

(a) Gondwanaland
(b) Laurasia
(c) Pangea
(d) Panthalasea

120. Mechanical erosion can not take place in following manner

(a) Abrasion
(b) Attrition
(c) Cavitation
(d) Dissoliving action

121. Principal of uniformitarian is proposed by

(a) Williem Smith
(b) James Hutton
(c) Gottlob Wormer
(d) Harry Hess

122. According to plate tectonics Theory, most active volcanoes occur

(a) On continents
(b) In large tectonic plates
(c) Along plate boundaries
(d) Randomly ore continents

123. Which one of the following is a type of volcanic eruption?

(a) Caledonian (b) Strombolian
(c) Venatian (d) Bombolian

124. Ventifacts are found in

(a) Deep Oceans (b) Glaciers
(c) Volcanic eruption (d) Deserts

125. A lagoonal circular coral reef is called

(a) Blue lagoon (b) Atoll
(c) Barrier type coral reef (d) Marginal reef

126. What can be expected at a depth of about 35 km inside the earth?

(a) Inner mantle
(b) Inner core
(c) Mohorovicic discontinuity
(d) Gutenberg discontinuity

127. The Richter scale measures

(a) Magnitude of earthquake
(b) Distance of location of earthquake
(c) Intensity of earthquake
(d) Depth of epicenter of earthquake

128. Tsunami is a

(a) Typhoon (b) Seismic sea wave
(c) Cyclone (d) None of the above

129. The term 'piracy' is associated with which of the following?

(a) Ocean (b) Wind
(c) River (d) Lake

130. Sediment transport in which particles are moved forward in series of short leaps and bounces is called

(a) Traction (b) Suspension
(c) Saltation (d) None of these

131. The average radius of the earth is

(a) 6378 km (b) 637.1 km
(c) 6372 km (d) 6271 km

132. 'Sima' is a part of

(a) Outer core (b) Crust
(c) Upper mantle (d) Lower mantle

133. The crust and upper part of upper mantle together constitute

(a) Troposphere (b) Asthenosphere
(c) Lithosphere (d) Biosphere

134. The instrument used for recording seismic waves is

(a) Thermometer (b) Seismograph
(c) Barometer (d) Seismogram

135. The 'Nebular Hypothesis' was proposed by

(a) Kant (b) Laplace
(c) Kant and Laplace (d) Moulten and Chamberlin

136. 'Conrad discontinuity' lies between

(a) Crust and Mantle (b) Sial and Sima
(c) Sima and Mantle (d) Mantle and Core

137. The upper layer of earth is known as

(a) Mantle (b) Core
(c) Crust (d) None of these

138. The continental drift theory was proposed by.

(a) Wegener (b) Pratt
(c) Hess (d) Hall

139. Most violent type of volcano is called

(a) Hawiian Type (b) Strombolian Type
(c) Vulcanian Type (d) Pelean Type

140. 'Sial' is a part of

(a) Outer core (b) Upper mantle
(c) Upper Crust (d) Lower mantle

141. The supercontinent that existed between 900-1000 million years ago is named as:

(a) Pangea (b) Panthalasa
(c) Rodinia (d) Gondwanaland

142. The volcanism at Iceland maybe:

(a) MOR related
(b) Hot Spot related
(c) Both MOR and Hot Spot related
(d) None of the above

143. Which of the following oceans is contracting?

(a) Indian Ocean (b) Atlantic Ocean
(c) Pacific Ocean (d) None of the above

144. Seismic continental margins are usually associated with

(a) Oceanic trench (b) Microcontinents
(c) Transform faults (d) Ocean ridge

145. The Mediterranean represents an ocean in the _______ stage of the life

(a) Embryonic (b) Mature
(c) Declining (d) Terminal

146. Mid-ocean ridge is formed due to the plate movement which is called

(a) Convergent movement
(b) Divergent movement
(c) Transform fault movement
(d) Lateral slipping plate movement

147. Where would deep - focus earthquakes occur?

(a) Under spreading centres
(b) At subduction zone

(c) Along the transform faults

(d) At Mid-ocean ridge

148. Which of the following pairs of plumes is associated with rifting and breakup events between India-Madagascar-Seychelles that led to the development of western continental margin of India?

(a) Reunion and Marion (b) Crozet and Marion

(c) Marion and Kerguelen (d) Reunion and Crozet

149. "Present is the key of past" is associated with which of the following principles of stratigraphy

(a) Law of superposition

(b) Principle of lateral continuity

(c) Principle of original horizontality

(d) Principle of uniformitarianism

150. Where do the ophiolite suit of rocks occur?

(a) Indus suture zone (b) Lesser Himalaya

(c) Brahmaputra valley (d) Himalayan foothill zone

151. Regions where continental lithosphere is currently undergoing extension or underwent extension in the past are termed as:

(a) Rifts (b) Subduction

(c) Obduction (d) Trench

152. Which one of the following margins represents plate boundaries?

(a) Active continental margins

(b) Passive continental margins

(c) Active ocean margins

(d) Passive ocean margins

153. Which one of the following backarcs is commonly called "Mariana- type backarcs"?

(a) Extensional backarcs (b) Compressional backarcs

(c) Twisting backarcs (d) Couple backarcs

154. The majority of Earth's Fe and Ni are found in:

(a) Lower crust (b) Upper crust

(c) Upper mantle (d) Inner core

155. Consider the following statements regarding continental crust and oceanic crust:

1. Continental crust is less dense and thicker than oceanic crust
2. Continental crust is poor in Iron and Magnesium
3. Continental crust is rich in Iron and Magnesium
4. Continental crust is more dense and thinner than oceanic crust

Which of the statements given above is/are correct?

(a) 1 only (b) 1 and 2
(c) 2 and 4 (d) 1 and 3

156. Which one of the following is the correct order of abundance of different elements in weight percentage in bulk Earth?

(a) O>Si>Fe>S>Ca>Al>Mg>Ni
(b) Fe>O>Si>Mg>S>Ni>Ca>Al
(c) Fe>Si>O>Mg>Ni>S>Ca>Al
(d) Si>Fe>O>Ni>S>Ca>Al>Mg

157. Which seismic wave is responsible for maximum shaking during an earthquake?

(a) P-wave (b) S-wave
(c) Rayleigh wave (d) Love wave

158. Bhuj earthquake of 2001 belonged to which one of the following categories?

(a) Deep focus
(b) Shallow focus
(c) Intermediate focus
(d) In between deep and intermediate focus

159. Which type of clay is found in the deepest parts of the ocean and is also abundant near the volcanoes?

(a) Black clay (b) Pelagic clay
(c) Grey clay (d) Green clay

160. Ocean tides are generated due to:

(a) Gravitational attraction of the moon
(b) Gravitational attraction of the sun

(c) Declination of the moon and the sun

(d) Gravitational force of the earth

Select the correct answer using the code given below

(a) 1,2, and 4 (b) 2,3 and 4

(c) 1,2, and 3 (d) 1,3, and 4

161. Volcanic island arcs are associated with __________.

(a) Transform plate boundaries

(b) Divergent plate boundaries

(c) Ocean-ocean convergent plate boundaries

(d) Ocean-continent convergent plate boundaries

162. The two types of crust are __________ and _______________.

(a) Oceanic, volcanic

(b) Oceanic, continental

(c) Continental, volcanic

(d) Oceanic, land

163. Iceland, the Samoan Islands, and Hawaii are examples of landforms created by _____.

(a) A convergent zone (b) A subduction zone

(c) A ring of fire (d) A hot spot

164. At mid-ocean ridges, new ___________crust is created.

(a) Volcanic (b) Transform

(c) Convergent (d) Oceanic

165. Seafloor spreading is a process that creates new ________ crust.

(a) Convergent (b) Oceanic

(c) Volcanic (d) Transform

166. The ________Islands are located in the middle of the Pacific plate.

(a) Icelandic (b) Samoan

(c) Australian (d) Hawaiian

167. Convection currents in our Earth's mantle drives the process that moves our _______.

(a) Mountains (b) Oceans

(c) Crusts (d) Volcanoes

168. The largest and most powerful mountains and volcanoes are located _______.

(a) On the surface (b) Underwater

(c) Above surface (d) On Iceland

169. Areas where continental and oceanic plates converge are also said to be ________.

(a) Mid-Pacific zones (b) Subduction zones

(c) Icelandic zones (d) Mid-Atlantic zones

170. Between oceanic and continental crusts, the ____________crust is denser.

(a) Continental (b) Oceanic

(c) Convergent (d) Divergent

171. The rate of seafloor spreading is approximately __________ every 10 years.

(a) 3 inches (b) 30 cm

(c) 30 inches (d) 3 cm

172. As the ocean seafloor spreads, the newly formed floor will adapt a specific magnetic signature according to the Earth's magnetic pole at that time. As years pass, lateral segments of seafloor will thus share the same _____signature on either side of the mid-oceanic ridge.

(a) Seductive (b) Divergent

(c) Convergent (d) Magneti

173. In seafloor spreading, the further the ocean floor is away from the mid-ocean ridge, the ___________that floor will be.

(a) Prettier (b) Newer

(c) Older (d) Younger

174. Iceland is located in the middle of ___________.

(a) The mid-Pacific ridge (b) The mid-convergent ridge

(c) The mid-Caribbean ridge (d) The mid –Atlantic ridge

175. In an island chain such as Hawaii, volcanic islands further away from the hot spot will tend to be ________.

(a) Older (b) Prettier

(c) Rockier (d) Newer

ANSWER SHEET

1. (b)	2. (d)	3. (c)	4. (d)	5. (d)
6. (b)	7. (a)	8. (c)	9. (b)	10. (a)
11. (b)	12. (a)	13. (d)	14. (c)	15. (d)
16. (d)	17. (d)	18. (c)	19. (c)	20. (a)
21. (b)	22. (d)	23. (d)	24. (a)	25. (c)
26. (c)	27. (a)	28. (a)	29. (b)	30. (b)
31. (c)	32. (b)	33. (c)	34. (d)	35. (c)
36. (d)	37. (b)	38. (c)	39. (b)	40. (a)
41. (c)	42. (a)	43. (b)	44. (a)	45. (b)
46. (c)	47. (c)	48. (d)	49. (c)	50. (a)
51. (b)	52. (a)	53. (b)	54. (b)	55. (d)
56. (b)	57. (b)	58. (c)	59. (b)	60. (b)
61. (a)	62. (c)	63. (b)	64. (d)	65. (a)
66. (a)	67. (d)	68. (c)	69. (a)	70. (b)
71. (c)	72. (d)	73. (a)	74. (d)	75. (c)
76. (b)	77. (a)	78. (d)	79. (b)	80. (b)
81. (c)	82. (c)	83. (d)	84. (d)	85. (d)
86. (c)	87. (a)	88. (a)	89. (d)	90. (a)
91. (d)	92. (c)	93. (b)	94. (a)	95. (a)
96. (d)	97. (a)	98. (b)	99. (b)	100. (c)
101. (b)	102. (d)	103. (b)	104. (a)	105. (b)
106. (b)	107. (d)	108. (b)	109. (b)	110. (c)
111. (b)	112. (b)	113. (c)	114. (a)	115. (d)
116. (d)	117. (a)	118. (b)	119. (c)	120. (d)
121. (b)	122. (c)	123. (b)	124. (d)	125. (b)
126. (c)	127. (a)	128. (b)	129. (c)	130. (c)
131. (b)	132. (b)	133. (c)	134. (b)	135. (c)
136. (b)	137. (c)	138. (a)	139. (d)	140. (c)
141. (c)	142. (c)	143. (c)	144. (a)	145. (d)
146. (b)	147. (b)	148. (a)	149. (d)	150. (a)
151. (a)	152. (a)	153. (a)	154. (d)	155. (b)
156. (b)	157. (d)	158. (b)	159. (b)	160. (c)
161. (c)	162. (b)	163. (d)	164. (d)	165. (b)
166. (d)	167. (c)	168. (b)	169. (b)	170. (b)
171. (b)	172. (d)	173. (c)	174. (d)	175. (a)

CHAPTER - 2

GEOMORPHOLOGY

GEOMORPHOLOGY

1. When O_2 <10power-6, that the environment is called:

(a) Oxic (b) Anoxic

(c) Subanoxic (d) Suboxic

2. SMOW (Standard Mean Ocean Water) is used in the calibration of ___isotopes.

(a) C (b) S

(c) N (d) O

3. Which one of the following drainage patterns is typical of a doubly plunging antiformal terrain?

(a) Dendritic (b) Trellis

(c) Rectangular (d) Radial

4. Which one of the following is most violent eruption?

(a) Hawaiian (b) Plinian

(c) Pelean (d) Strombolian

5. In the Himalayan mountain belt one can find abundant

(a) Normal fault and fold

(b) Thrust and fold

(c) Strike slip fault and bending of Rick strata

(d) Gravity induced sag structure

6. The Mohorovicic discontinuity below the Himalayan orogen is at a depth greater than that below the Indian craton. This is because of

(a) Isostatic mass adjustment

(b) Lateral density variation

(c) Upwelling material below the Indian craton

(d) Less tectonic activity in the Himalayan region

7. The Environmental system that is not affected by human interference is known as:

(a) Natural system (b) Ecosystem

(c) Controlled system (d) Closed system

8. The study of history of an organism from its death to the pre-burial stage is known as:

(a) Ecology (b) Digenesis

(c) Morphogenesis (d) Biostratinomy

9. Which of the following geomorphic suggest regional uplift during the Quaternary?

(a) Lateral moraines (b) Paired terraces

(c) Recessional moraines (d) Unpaired terraces

10. Which of the following geomorphic feature is not an indicator of active tectonic:

(a) Incised valleys (b) River terraces

(c) Stream piracy (d) U-shaped valley

11. Deccan Basalt represents:

(a) A Plateau (b) A Mountain chain

(c) A Hillock (d) A Basin

12. Fluvial environment is a:

(a) Continental environment (b) Marine environment

(c) Transitional environment (d) Abyssal environment

13. Shallowing upward sequence indicate:

(a) Marine transgression (b) Marine regression

(c) High stand (d) Stagnant Sea level

14. Which of the following problem is not of anthropogenic origin?

(a) Landslide donation (b) Groundwater pollution

(c) Flash food (d) Air pollution

15. Tidal flats are

(a) Shore line environment (b) Deep sea environment

(c) Glacial environment (d) Lacustrine environment

16. A Hanging valley is formed by-

(a) River (b) Wind

(c) Glacier (d) Ocean

17. Oasis are formed in deserts due to the process of

(a) Deflation (b) Abrasion

(c) Attrition (d) Evaporation

18. The maximum advancement of a glacier is depicted by the position of the

(a) Medical moraine (b) Terminal moraine

(c) Lateral moraines (d) Ground moraines

19. Life on Earth originated at microscopic level under

(a) Oxidising environment

(b) Reducing environment

(c) Extremely high environment

(d) Extremely cold environment

20. Which of the following geomorphic feature indicates rejuvenation

(a) Barchan Dune (b) Meandering river channel

(c) Paired River terraces (d) Broad flood plains

21. Baikal lake has formed due to the process of

(a) Rifting (b) Glaciation

(c) Solution activity (d) Meteoritic impact

22. Ecosystem is the study of

(a) Biological and physical systems

(b) Biological systems

(c) Physical systems

(d) Geological systems

23. The sum total of all the feature and conditions that surround an organism is the

(a) Atmosphere (b) Hydrosphere

(c) Biosphere (d) Environment

24. Village formed by glaciers always have which of the following characteristics?

(a) U-saped cross-section

(b) Straight course

(c) Uniform slope

(d) Bedrock floors

25. The curved body of water formative when a meandering river reshapes and cuts off one of its bends is called:

(a) An Essing pond (b) An Oxbow lake

(c) A Swensen pool (d) A Groovy pool

26. The budgets that form win the eyes in a glassier melts are called:

(a) Drifts (b) Dunes

(c) Venus (d) Cirques

27. A circular cavity ground out of bedrock by running water is called a:

(a) Cave (b) Pot hole

(c) Sink hole (d) Kettle hole

28. What is the flow pattern of a River system that resembles the tree branches?

(a) Trellis pattern (b) Dendritic pattern

(c) Radial pattern (d) Rectangular pattern

29. Which of the following types of rivers receive their discharge from the groundwater of the area?

(a) Effluent rivers (b) Intermittent rivers

(c) Perennial rivers (d) Influent rivers

30. Which of the following agents from Roches Moutonees?

(a) Wind (b) River

(c) Sea (d) Glacier

31. Match List-I with List-II of Aeolian landforms and select the correct answer using the code given below the Lists:

List-I	List-II
(A) Interior basin surrounded by mountain	1. Playa
(B) Talus apron at the foot of the mountain	2. Pediment

(C) Upland with lag gravel	3. Hammada
(D) Seasonal salt pan at the centre of the basin	4. Bolson

Code:

	A	B	C	D
(a)	1	3	2	4
(b)	4	3	2	1
(c)	1	2	3	4
(d)	4	2	3	1

32. Which one of the following processes results in development of an oasis?

(a) By the filling in of a depression by water

(b) By the meteoric fall causing a depression on the ground

(c) By wind erosion down to water level in a desert

(d) When an advancing glacier terminates

33. Match List-I with List-II and select the correct answer using the code given below the Lists:

List-I	**List-II**
(A) Barchans	1. $CaCO_3$ Columnar deposits on the floor of a cave
(B) Wave-built terrace	2. Formed by river action
(C) Stalagmite	3. An Aeolian deposit
(D) Cascade	4. Feature of a Sea-profile

Code:

	A	B	C	D
(a)	2	4	1	3
(b)	3	4	1	2
(c)	2	1	4	3
(d)	3	1	4	2

34. Roches Moutonnee land form develops due to

(a) Glacial Erosion (b) Wind Erosion

(c) Ground Water Erosion (d) River Erosion

35. Loess land form is result of -

(a) Glacial Erosion (b) Glacial Deposition

(c) Wind Deposition (d) Wind Erosion

36. Karst topography develops in which region -

(a) Granitic region (b) Carbonate region

(c) Basaltic region (d) Sandy desert

37. Stalactites are present in which place -

(a) Carbonate cave (b) Pond deposit

(c) Desert topography (d) River valley

38. Tillite rock forms due to -

(a) Wind Action (b) Ground Water Erosion

(c) Oceanic water work (d) Glacial work

39. Moraine develops due to -

(a) Geological work of River

(b) Geological work of Wind

(c) Geological work of Glacial work

(d) Geological work of Ground water

40. Exfoliation results due to -

(a) Mechanical Weathering (b) Chemical decay

(c) Wind abrasion (d) Glacial movement

41. Bauxite deposit develops due to -

(a) Glacial erosion

(b) Wind erosion and deposition

(c) Chemical weathering and Solution process

(d) Frost Action

42. TheArkosehas dominance which of the following minerals?

(a) Calcite and Dolomite (b) Clay minerals

(c) Quarts and Feldspar (d) Garnet and Magnetite

43. "Tuff" is an Example of which type of deposit -

(a) Glacial Deposit (b) Aeolian Deposit

(c) River Deposit (d) Pyroclastic Deposit

44. Which of the following is the second highest layer of the Earth Atmosphere -

(a) Stratosphere (b) Mesosphere

(c) Troposphere (d) Inosphere

45. Which among in the following region in India receives the least average annual monsoon rainfall -

(a) Jaipur (b) Laddakh

(c) Shillong (d) Nagpur

46. The EI-NINO happens became -

(a) Trade winds weaken and cannot bring the cold water

(b) Trade winds make the water cold

(c) Trade wind strengthen and cannot bring up cold water

(d) Trade winds strengthen and bring up the warm water

47. Which one of the following action forms the pedestal rock?

(a) Wind action (b) Glacial action

(c) River action (d) Sea-water action

48. Match List-I with List-II and select the correct answer using the codes given below the

Lists:

List-I	List-II
A. Well sorted sand	1. Glacial deposit
B. Dropstone	2. Beach deposit
C. Graded bed	3. Turbidite deposit
	4. Fluvial deposit

Code:

	A	B	C
(a)	3	1	4
(b)	2	1	3
(c)	4	2	1
(d)	2	3	4

49. Which one is not a river pattern?

(a) Knick (b) Escarpment

(c) Mesa (d) Peneplanis

50. Which one is not a river pattern?

(a) Antecedent (b) Consequent
(c) In sequent (d) Trellis

51. Which of the following term is not related to wind erosion?

(a) Deflection (b) Traction
(c) Abrasion (d) Attrition

52. Blow-outs formed due to

(a) Wind (b) River
(c) Glacier (d) Impact

53. Fiords are formed by action of

(a) Wind (b) River
(c) Glacier (d) Lake

54. Which of the following is not regional process of continental Glaciers?

(a) Plucking (b) Avalanching
(c) Rasping (d) None of the above

55. Caverns are related to

(a) Underground water (b) River
(c) Glacier (d) Wind

56. Blind valley is formed due to

(a) Erosional (b) Deposition
(c) Steep slope (d) Surface runoff

57. In the following, which combination is wrong?

(a) Knick point, Escarpment, Mesa
(b) Plucking, Avalanching, Rasping
(c) Hamada, Yardang, Sand dunes
(d) Pedestal, Cirques, Arête

58. The term caldera is associated with

(a) River (b) Volcano
(c) Glacier (d) Wind

59. A V-shaped valley is characteristic of

(a) Glacier (b) Ocean
(c) River (d) Lake

60. Which one of the following is not a glacial feature?

(a) Yardang (b) Bergschrund
(c) Arete (d) Crevasse

61. Seifs are found in which environments?

(a) Glacial (b) Fluvial
(c) Lacustrine (d) Desert

62. Rejuvenation in an area is suggested by which of the following?

(a) Waterfalls (b) Paired river terraces
(c) V-shaped valley (d) All of the above

63. Thermohaline circulation results due to difference in

(a) Density (b) Temperature
(c) Salinity (d) All of these

64. Which of the following marks the farthest advance of a glacier?

(a) Terminal moraine (b) Lateral moraine
(c) Medial moraine (d) Ground moraine

65. Long, narrow and sinuous ridges of sands and gravels

(a) Drumlins (b) Crag and tail
(c) Eskers (d) Kames

66. Which among the following is formed by wind erosion?

(a) Yardang (b) Gorges
(c) Loess (d) Butte

67. Headlands are produced by

(a) Groundwater erosion (b) River erosion
(c) Marine erosion (d) Wind erosion

68. Stalagmites are characteristic feature of:

(a) River (b) Glacial
(c) Groundwater (d) Wind

69. Which among the following is a feature produced by wind?

(a) Drumlins (b) Loess

(c) Delta (d) Canyons

70. Antidunes develop under

(a) Transitional flow regime

(b) Lower flow regime

(c) Upper flow regime

(d) Both in lower and upper flow regime

71. Bouma sequence forms in

(a) Shallow sea (b) Rivers

(c) Desert (d) Deep sea

72. Average depth of fair weather wave base is

(a) 40-50 m (b) 8-15 m

(c) 70-80 m (d) 100 m

73. In shallow marine environment, palacocurrent patterns are

(a) Unimodal (b) Bimodal

(c) Polymodal (d) Both unimodal and bimodal

74. Braided River Channels develop due to

(a) Low surface gradient and stable banks

(b) High suspension load

(c) Steep surface gradient and unstable banks

(d) Point bars

75. Delta is formed by

(a) Wind (b) River

(c) Glacier (d) All of these

76. Dune is formed by

(a) Wind (b) Glacier

(c) River (d) None of these

77. The environment between the high tide and low tide levels of the sea is known as

(a) Neritic (b) Bathyal

(c) Abyssal (d) Littoral

78. Sinkhole, doline and drip stone are significant features of

(a) Aeolian landform (b) Karst topography

(c) Glacial landform (d) Fluvial landform

79. Which of the following is formed by wind erosion

(a) Gorges (b) Loess

(c) Yardang (d) Butte

80. Which one of the following geomorphic features can be formed by erosion

(a) Hook (b) Terrace

(c) Loess (d) Stalactite

81. Headlands are produced by

(a) Ground water erosion (b) River erosion

(c) Marine erosion (d) Wind erosion

82. Which one among the following is a feature produced by wind

(a) Drumlins (b) Loess

(c) Delta (d) Canyons

83. The mature stage feature of the river is

(a) Alluvial fan (b) V-shaped valley

(c) Meanders (d) Ox-bow lake

84. Deltaic deposits are characterised by

(a) Coarsening upward sediment cycles

(b) Fining upward sediment cycles

(c) Mixed cycles

(d) None of the above

85. Tidal bundles form in

(a) Rivers (b) Glaciers

(c) Shallow marine environment (d) Mountains

86. The process in which the glacial wastage takes place by the double process of evaporation and melting is known as

(a) Nivation (b) Calvation

(c) Ablation (d) Plucking

87. Which one of the following minerals has highest susceptibility to weathering?

(a) Olivine (b) Pyroxene
(c) Muscovite (d) Quartz

88. The ridge like deposits of silt and clay along the borders of the flooded rivers after a flood are known as:

(a) Placer deposit (b) Natural levees
(c) Scree (d) Cuesta

89. Which one of the following is NOT a sink hole?

(a) Uvala (b) Polje
(c) Solution pan (d) Lapies

90. Which one of the following is NOT related to glaciers?

(a) Doline (b) Firn
(c) Cirque (d) Drumlins

91. A graph showing percentage of Earth's surface area as a function of elevation or depth relative to sea level is called:

(a) Hypsometric curve (b) Stress curve
(c) Strain curve (d) Creep curve

92. Mushroom, inselberg and zeugen are produced by:

(a) Wind erosion (b) Glacial erosion
(c) River erosion (d) Sea erosion

93. Paternoster lake is characteristic of:

(a) Aeolian action (b) Glacial action
(c) Fluvial action (d) Stream terraces

94. Alluvial fans are significant:

(a) Fluvial landforms (b) Glacial landforms
(c) Aeolian landforms (d) Karst topography

95. Long, narrow and sinuous ridges of sands and gravels situated in the middle of ground moraines are:

(a) Drumlins (b) Crag and tail
(c) Eskers (d) Kames

96. Which of the following is formed by wind erosion?

(a) Yardang (b) Gorges

(c) Loess (d) Butte

97. Headlands are produced by:

(a) Groundwater erosion (b) River erosion

(c) Marine erosion (d) Wind erosion

98. Stalagmites are characteristic feature of:

(a) River (b) Glacier

(c) Groundwater (d) Wind

99. Which one among the following is a feature produced by wind?

(a) Drumlins (b) Loess

(c) Delta (d) Canyons

100. A lagoonal circular coral reef is called:

(a) Blue lagoon (b) A toll

(c) Barrier type coral reef (d) Marginal reef

ANSWER SHEET

1.	(b)	2.	(d)	3.	(d)	4.	(c)	5.	(b)
6.	(a)	7.	(a)	8.	(d)	9.	(b)	10.	(d)
11.	(a)	12.	(a)	13.	(b)	14.	(c)	15.	(a)
16.	(c)	17.	(a)	18.	(b)	19.	(b)	20.	(c)
21.	(a)	22.	(d)	23.	(d)	24.	(a)	25	(b)
26.	(a)	27	(b)	28.	(b)	29	(a)	30.	(d)
31.	(b)	32.	(c)	33.	(b)	34.	(a)	35.	(c)
36.	(b)	37.	(a)	38.	(d)	39.	(c)	40.	(a)
41.	(c)	42.	(c)	43.	(d)	44.	(b)	45.	(b)
46.	(a)	47.	(a)	48.	(a)	49.	(c)	50.	(d)
51.	(b)	52.	(a)	53.	(c)	54.	(b)	55.	(a)
56.	(a)	57.	(d)	58.	(b)	59.	(c)	60.	(a)
61.	(d)	62.	(b)	63.	(d)	64.	(a)	65.	(c)
66.	(a)	67.	(c)	68.	(c)	69.	(b)	70.	(c)
71.	(d)	72.	(b)	73.	(c)	74.	(d)	75.	(b)
76.	(a)	77.	(d)	78.	(b)	79.	(c)	80.	(b)
81.	(c)	82.	(b)	83.	(d)	84.	(a)	85.	(c)
86.	(c)	87.	(a)	88.	(b)	89.	(c)	90.	(a)
91.	(a)	92.	(a)	93.	(b)	94.	(a)	95.	(c)
96.	(a)	97.	(c)	98.	(c)	99.	(b)	100.	(c)

CHAPTER - 3

MINERALOGY

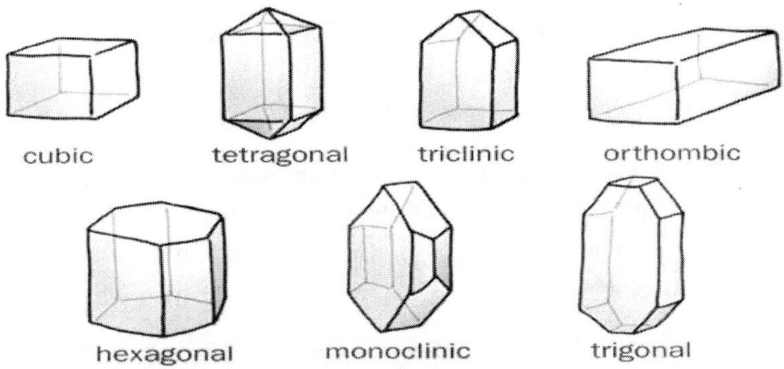

MINERALOGY

1. **Which of the following agents is responsible for mechanical loosening and removal of material by deflation?**
   (a) River (b) Wind
   (c) Glacier (d) Rain-water

2. **Crystal forms that neither have plane nor have centre of symmetry are called _____forms.**
   (a) Hemihedral (b) Hemimorphic
   (c) Enantiomorphic (d) Tetrahedral

3. **The angle subtended between two optic axes is called:**
   (a) 2/2V (b) 2V
   (c) 1/2V (d) 1/4V

4. **When crystallographic axes and compositional plane in feldspar is basal pinacoid (001), then the twin is called:**
   (a) Mane Bach twin (b) Carlsbad twin
   (c) Boven twin (d) Hour Glass twin

5. **What is the refractive index of a Canada balsam?**
   (a) 1.34 (b) 1.44
   (c) 1.54 (d) 1.64

6. **The sign of elongation in crystals is determined by:**
   (a) Mica plane (b) Break compensator
   (c) Gypsum plate (d) Quartz wedge

7. **Which of the following minerals belongs to a common Galena type?**
   (a) Ilmenite (b) Spinel
   (c) Boracite (d) Sphalerite

8. **For olivine, common effects of dispersion in Biaxial minerals is the dispersion of the optic axes and is expressed as ______ dispersion**
   (a) Feeble (b) Large
   (c) Small (d) Weak

9. Which of the following pyroxenes in identified by smallest 2V?

(a) Enstatite (b) Hypersthene

(c) Diopside (d) Pigeonite

10. In which of the following minerals, straight extinction is observed?

(a) Nosean (b) Garnet

(c) Augite (d) Biotite

11. The angle between two sets of cleavage planes in hornblende is nearly:

(a) 30° (b) 60°

(c) 90° (d) 120°

12. Which of the following minerals is used in the removal of heavy metals from nuclear, mine and industrial waste water?

(a) Analcime (b) Periclase

(c) Plagioclase (d) Orthoclase

13. Which of the following minerals shows internal reflection with its diagnostic colour?

(a) Sphalerite (b) Covellite

(c) Pyrrhotite (d) Pyrite

14. Which of the following pyroxenes is Biaxial negative?

(a) Enstatite (b) Augite

(c) Hypersthene (d) Diopside

15. In olivine structure, the layers consisting of octahedral are cross-linked to independent SiO_4 tetrahedra that lie parallel to which of the following?

(a) (010) (b) (100)

(c) (001) (d) (111)

16. What is the Si : O ratio in phyllosilicates?

(a) 1 : 3 (b) 1 : 4

(c) 4 : 10 (d) 6 : 12

17. Tectosilicates represent Silicon: Oxygen ratio of:

(a) 1 : 2 (b) 1 : 3

(c) 1 : 4 (d) 2 : 5

18. Basal section of Uniaxial crystal when viewed under conoscopic conditions show:

(a) Emergence of optic axis (b) Eccentric isogyres

(c) Uncentred isogyres (d) Perfectly cantered isogyres

19. Hornblende shows:

(a) 1st order polarization colours

(b) 2nd order polarization colours

(c) 3rd order polarization colours

(d) 4th order polarization colours

20. Which of the following minerals show pink, yellow and green pleochroic colors along X, Y and Z axes?

(a) Augite (b) Hornblende

(c) Biotite (d) Hypersthene

21. What are the correct indices of 1a : 1b : 1c faces?

(a) HKL (b) HLL

(c) HLO (d) HHL

22. When the crystallographic axis is fast, the mineral is said to be:

(a) Uniaxial optically negative (b) Uniaxial optically positive

(c) Biaxial optically negative (d) Biaxial optically positive

23. By which of the following groups of elements are alkaline rocks characterised?

(a) Li-Be-B (b) Fe-Mg-Ti

(c) Mg-Fr-Cr (d) Si-K-Na

24. Which of the following siliciclastic end members demonstrate increased diagenetic?

(a) Quartz (b) Orthoclase

(c) Muscovite (d) Plagioclase

25. The characteristics pyroxene found in eclogite is:

(a) Omphacite (b) Pigeonite
(c) Diopside (d) Hedenbergite

26. A very high K basaltic andesite is called:

(a) Abenakiite (b) Boninite
(c) Adakite (d) Shoshonite

27. Which of the following is known as a High Field Strength Element in geochemistry?

(a) Zirconium (b) Barium
(c) Strontium (d) Rubidium

28. Which of the following mineral pairs does not represent polymorphism?

(a) Calcite-Aragonite (b) Rutile-Brookite
(c) Albite-Anorthite (d) Graphite-Diamond

29. Twinning observed in plagioclase is known as –

(a) Polysynthetic twinning (b) Carlsbad twinning
(c) Penetration twinning (d) None of the above

30. Which of the following gemstones would be smallest in size if each weigh 10 carats?

(a) Diamonds (b) Garnet
(c) Topaz (d) Zircon

31. Which of the following minerals is different from others in chemical composition?

(a) Coe site (b) Celestite
(c) Stishovite (d) Chalcedony

32. Which of the following represents the correctly the Bragg's equation?

(a) $n= 2d \sin\Theta$ (b) $n\lambda= 2d \sin\Theta$
(c) $2n = d\lambda\sin\Theta$ (d) $\lambda = 2d \sin\Theta$

33. Which one of the following gemstones is organic in origin?

(a) Jadeite (b) Diamond
(c) Jet (d) Opal

34. Malachite and Azurite are characteristics ore minerals of -

(a) Gossan Zone

(b) Zone of oxidation

(c) Zone of supergene enrichment

(d) Proctor

35. The minerals Kyanite exhibit a variation in the hardness from 4.5 to 7, in different directions. The highest value is found parallel to which of the following axis?

(a) C- Axis (b) B-Axis

(c) A-Axis (d) B and C-Axis

36. Which of the following method is used to determine the space lattice structure of mineral?

(a) Chemical analysis (b) Spectroscopic method

(c) X-Ray diffraction (d) Scanning electron microscope

37. In Moh's Scale of hardness, how many minerals are of silicate composition?

(a) 4 (b) 5

(c) 6 (d) 7

38. Which of the following minerals is most resistant to weathering?

(a) Quartz (b) Muscovite

(c) Olivine (d) Sodic plagioclase

39. Between crossed Polaris, a colourless mineral show I order grey interference colours, but is not very distinct in plane polarised light. The refractive index of the mineral should be around:

(a) 1.0082 (b) 1.2495

(c) 1.5356 (d) 1.7221

40. Which of the following instrument can be used to identify quartz and microcline in a rock sample?

(a) Petrological microscope and XRF

(b) Petrological microscope and XRF

(c) Petrological microscope and UV-VIS spectrophotometer

(d) Petrological microscope and Flame photometer

41. In a mineral the angle between the cleavage traces and one of the vibration direction is 90°. The mineral show which type of extinction?

(a) Parallel (b) Inclined

(c) Symmetrical (d) Wavy

42. In a petrological microscope the optical tube commonly contain.

(a) Objective lens, Condenser lens, Polariser lens, Bertrand lens

(b) Eye piece, Analyser, Bertrand lens, Accessory plate slot

(c) Eye piece, Analyser, Diaphragm

(d) Mirror, Polariser, Condenser lens, Diaphragm

43. A mineral belonging to the Hexagonal system will show:

(a) Interference colour; uniaxial interference figure and large 2V

(b) Interference colours; behave sometimes as isotropic and show uniaxial interference figure

(c) Always isotropic

(d) Bright colours

44. In biaxial minerals the intermediate refractive index is always along

(a) X vibration direction (b) Y vibration direction

(c) Z vibration direction (d) A crystallographic axis

45. The mineral orthoclase and microcline can be differentiated with the help of which instrument?

(a) SEM (b) Flame photometer

(c) TEM (d) X-Ray diffractometric

46. Which of the following mineral property cannot be seen between crossed polars

(a) Cleavage (b) Pleochroism

(c) Inclusions (d) Form

47. In the degree of inclined extinction, the vibration directions of the mineral make an angle of ......with the vibration directions of the Polaroids.

(a) Zero degree (b) Some degree

(c) Forty degree (d) Ninety degree

48. The interference colour of a mineral seen in petrological microscope depends upon it's

(a) Crystal system and Pleochroism
(b) Birefringence and thickness
(c) Extinction type
(d) Twinning and extinction

49. Which of the following mineral is pleochroic in thin section?

(a) Quartz (b) Garnet
(c) Biotite (d) Plagioclase

50. One optic axis exists in minerals crystallising under which systems

(a) Cubic (b) Tetragonal
(c) Orthorhombic (d) Monoclinic

51. Which mineral property cannot be observed with a petrological microscope?

(a) Lustre (b) Twinning
(c) Extinction (d) Pleochroism

52. Tetrahedron is the hemihedral form of

(a) Octahedron (b) Rhombohedrum
(c) Tetrahedron (d) Rhombo-dodecahedron

53. Which mineral has no cleavage?

(a) Calcite (b) Garnet
(c) Mica (d) Kyanite

54. In minerals of tetragonal systems, there is no double refraction along which direction?

(a) Along c-axis (b) Along b-axis
(c) Along a-axis (d) None of these

55. On the basis of crystallographic axes, the crystal are divided into

(a) Two systems (b) Four systems
(c) Six systems (d) Eight systems

56. The refractive indices of a mineral are 1.258, 1.559 and 1.742. The mineral is –

(a) Biaxial (b) Uniaxial

(c) Can be Biaxial Uniaxial (d) Isotropic

57. In geological studies, Arkose contains which of the following minerals which gives it a reddish colour?

(a) Orthoclase (b) Calcite

(c) Quartz (d) Aragonite

58. The internal structure of crystals was first studied with which of the following?

(a) X-rays (b) Isotopes

(c) Electron microscopes (d) Lasers

59. Give the term that describes different crystalline modifications of the same chemical substance

(a) Polymorphism (b) Isomorphism

(c) Tetrahedral (d) Perovskite

60. Most natural mineral can be found in which two crystal system?

(a) Cubic and triclinic (b) Orthorhombic and monoclinic

(c) Hexagonal and tetragonal (d) Monoclinic and triclinic

61. Characteristics property of crystal is:

(a) Presence of face around in

(b) Transparency

(c) Presence

(d) Ordered atomic/molecular arrangement

62. Flat surface found around a crystal is called

(a) Cleavage (b) Joints

(c) Face (d) Plane

63. How many horizontal crystallographic axes are there in hexagonal system?

(a) 6 (b) 3

(c) 4 (d) 2

64. Into how many systems the crystals have been divided?

(a) 6 (b) 9

(c) 5 (d) 12

65. In Monoclinic system of crystals, the angle between which of the two axes is not 90°?

(a) A and C (b) A and B

(c) B and C (d) None of the above

66. In triclinic system of crystals, the angle between which of the two axes is 90°?

(a) A and C (b) A and B

(c) B and C (d) None of the above

67. Euhedral crystal have

(a) Fractures inside the crystals (b) Well developed faces

(c) Rounded corners (d) Poorly developed faces

68. What is the basis of dividing the crystal system into classes?

(a) Difference in the size of crystals

(b) Difference in symmetry elements present in crystals

(c) Presence or absence of cleavages

(d) None of the above

69. The size of crystals in cryptocrystalline material is

(a) Very large (b) Medium

(c) Microscopic (d) Small

70. In which of the following crystal system, two of the crystallographic axes are of same length?

(a) Cubic system (b) Orthorhombic system

(c) Tetragonal system (d) Monoclinic system

71. Gypsum type symmetry class belongs to which crystal system?

(a) Monoclinic system (b) Orthorhombic system

(c) Tetragonal system (d) Cubic system

72. Calcite type symmetry class belongs to which crystal system?

(a) Monoclinic system (b) Orthorhombic system

(c) Hexagonal system (d) Cubic system

73. Crystals of pyrite and tetrahedrite belong to which crystal system?

(a) Monoclinic system (b) Orthorhombic system

(c) Hexagonal system (d) Cubic system

74. Which type of symmetry axis is not found in crystals?

(a) 2-fold (b) 3-fold

(c) 5-fold (d) 6-fold

75. Interpenetration type of twinning is found in crystals of

(a) Pyrite (b) Muscovite

(c) Quartz (d) Tourmaline

76. Contact Goniometer is used to determine the

(a) Number of symmetry planes in the crystal

(b) Type of symmetry axis in the crystal

(c) Angle between two adjacent faces of the crystal

(d) Ratio of lengths of various crystallographic axes in the crystal

77. Which mineral is the most sense?

(a) Hematite (b) Cinnabar

(c) Marcasite (d) Rhodochrosite

78. Which of the following minerals is a source of lead?

(a) Bauxite (b) Hematite

(c) Malachite (d) Galena

79. Which of the following is the mineral name for ordinary salt?

(a) Biotite (b) Calcite

(c) Halite (d) Dolomite

80. Which of the following reasons explains why quartz cannot be identified by it's coloured one?

(a) Its colour is too faint

(b) It has no colour

(c) It exists in a variety of colours

(d) It's colour fades

81. The nickname, "Fool's Gold" refers to which of the following?

(a) Calcite (b) Apatite

(c) Fluorite (d) Pyrite

82. Which of the following is the chemical formula for calcite?

(a) $CaCO_3$ (b) $CaMgSO_4$

(c) $FeCO_3$ (d) FeS_2

83. Diamond and graphite are polymorphism of which of the following elements?

(a) Magnesium (b) Carbon

(c) Sulfur (d) Nitrogen

84. Which of the following pairs of minerals are difficult to tell apart because both are always dark in colour and almost equally hard?

(a) Muscovite and orthoclase (b) Muscovite and quartz

(a) Hornblende and talc (d) Hornblende and augite

85. Polysynthetic twinning is found in

(a) Orthoclase (b) Muscovite

(c) Augite (d) Plagioclase feldspar

86. Fluorite is at what number on Moh's hardness scale?

(a) 1 (b) 2

(c) 3 (d) 4

87. Which one of the following minerals does NOT contain silica tetrahedrons?

(a) Quartz (b) Muscovite

(c) Halite (d) Orthoclase

88. Which of the following minerals in noted its one perfect cleavage?

(a) Calcite (b) Muscovite

(c) Quartz (d) Pyrite

89. Which of the following minerals is NOT a silicate?

(a) Orthoclase (b) Muscovite

(c) Augite (d) Magnetite

90. The mineral halite is :

(a) Potassium chloride
(b) Sodium chloride
(c) Calcium chloride
(d) Calcium bromide

91. Agates is a form of what mineral?

(a) Magnetite
(b) Feldspar
(c) Quartz
(d) Copper

92. Which of the following minerals is composed of carbon?

(a) Calcite
(b) Baryte
(c) Diamond
(d) Gypsum

93. Which of the following mineral is isotropic?

(a) Quartz
(b) Beryl
(c) Tourmaline
(d) Garnet

94. The mineral generally used for construction of Nicole prism is:

(a) Dolomite
(b) Calcite
(c) Baryte
(d) Hornblende

95. How many Nicol prism are fitted in a petrological microscope?

(a) 2
(b) 4
(c) 1
(d) 3

96. How many carbonate minerals are in Moh's scale hardness?

(a) 2
(b) 1
(c) 4
(d) 3

97. The fundamental unit in building of silicate minerals is

(a) SiO_2
(b) SiO
(c) SiO_3
(d) SiO_4

98. Which of the following minerals does not show pleochroism?

(a) Biotite
(b) Glaucophane
(c) Hornblende
(d) Quartz

99. Which mineral has been assigned hardness of 7 in Moh's scale of hardness?

(a) Quartz
(b) Topaz
(c) Feldspar
(d) Apatite

100. The Si:O ratio in nesosilicate minerals is:

(a) 1 : 2 (b) 2 : 3

(c) 1 : 4 (d) 1 : 1

101. The lustre of quartz mineral is

(a) Pearly (b) Silky

(c) Greasy (d) Vitreous

102. Calcite is characterised by which of the following shape?

(a) Cubic (b) Fibrous

(c) Prismatic (d) Rhombic

103. Which of the following minerals does not have cleavages?

(a) Kyanite (b) Olivine

(c) Feldspar (d) Quartz

104. Which of the following minerals is not calcium rich?

(a) Calcite (b) Baryte

(c) Dolomite (d) Aragonite

105. Which of the following minerals is not composed of Al_2SiO_5?

(a) Andalusite (b) Sillimanite

(c) Kyanite (d) Siderite

106. Which of the following minerals is representative of cyclosilicate structure?

(a) Quartz (b) Olivine

(c) Beryl (d) Feldspar

107. Which of the following minerals is NOT a silicate?

(a) Orthoclase (b) Muscovite

(c) Augite (d) Magnetite

108. Which is Phyllosilicate mineral amongst the following -

(a) Olivine (b) Kyanite

(c) Quartz (d) Mica

109. The Si and O ratio in Soro silicate is -

(a) $Si_4 O_{10}$ (b) $Si_4 O_8$

(c) $Si_2 O_7$ (d) $Si_6 O_4$

110. Which mineral is Tecto silicate mineral -

(a) Augite (b) Hornblende

(c) Biotite (d) Quartz

111. The double chain silicate structure occurred in which mineral -

(a) Feldspar (b) Olivine

(c) Amphibole (d) Wollastonite

112. Which of the following is a high-pressure variety of silica?

(a) Chert (b) Opal

(c) Stishovite (d) Quartz

113. The silica mineral, which develops in high temperature acid volcanic is named as -

(a) Coesite (b) Tridymite

(c) Agate (d) Amethist

114. Generally the form of fluorite crystal is of -

(a) Tetrahedron Shape (b) Diploid Shape

(c) Octahedron Shape (d) Pyramid of first order

115. The Diamond Crystallises in -

(a) Triclinic System (b) Monoclinic System

(c) Isometric System (d) Hexagonal System

116. Gypsum crystals are classified in which system -

(a) Triclinic System (b) Orthorhombic System

(c) Cubic System (d) Monoclinic System

117. Which one of the following statements is correct? Oblique lattice invariant under rotation of

(a) $2\pi/3$ and π (b) π and 2π

(c) $2\pi/4$ and 2π (d) $2\pi/6$ and π

118. Which of the following ions occupy the octahedral sites in spinel?

(a) Monovalent ion (b) Bivalent ion

(c) Trivalent ion (d) Tetravalent ion

119. Which one of the following minerals belongs to Crystal class 1?

(a) Calcite (b) Albite
(c) Fluorite (d) Quartz

120. Which of the following symmetry elements characterize pyrite crystal?

(a) One sixfold axis (b) Two fourfold axis
(c) Three twofold axis (d) Four twofold axis

121. Which one of the following minerals does not exhibit pleochroism?

(a) Biotite (b) Hornblende
(c) Garnet (d) Chlorite

122. Which of the following is bounded by twentyfour similar faces, each meeting the axis at unequal distances and the general symbol being (hkl)?

(a) Dodecahedron (b) Diploid
(c) Tetrahedron (d) Pyritohedron

123. Which of the following crystals have basal pinacoid as a form?

(a) Rutile (b) Idocrase
(c) Zircon (d) Apophyllite

124. Select the correct answer from the code given below:

(a) 1 and 2 only (b) 2 and 4 only
(c) 3 and 4 only (d) 2, 3 and 4

125. Which of the following form is characterised by three or more nonparallel faces intersecting at a common point?

(a) Prism (b) Pinacoid
(c) Pyramid (d) Spheroid

126. Which one of the following minerals show strong pleochroism?

(a) Quartz (b) Muscovite
(c) Orthoclase (d) Hornblende

127. Which of the following types of extinction will be exhibited by the basal section of a monoclinic mineral under microscope?

(a) Straight (b) Parallel
(c) Symmetrical (d) Undulatory

128. Which one of the following minerals shows parallel extinction?

(a) Augite (b) Hypersthene

(c) Diopside (d) Hornblende

129. Which one of the following has highest refractive index?

(a) Sodalite (b) Fluorite

(c) Leucite (d) Pyrope

130. The isogyres would show which of the following position in case of large 2 V?

(a) Highly curved (b) At right angles to each other

(c) Touching each other (d) Virtually straight

131. In which of the following situation a biaxial mineral is positive?

(a) X is acute bisectrix (b) Y is acute bisectrix

(c) Z is acute bisectrix (d) Z is obtuse bisectrix

132. Which mineral occurs in small, hexagonal, colourless to white, plate like crystals found in cavities of volcanic rocks?

(a) Opal (b) Chalcedony

(c) Tridymite (d) Cristobalite

133. Which one of the following pairs is not correctly matched? (Mineral) (Moh's scale of hardness)

(a) Apatite : 5

(b) Gypsum : 2

(c) Orthoclase : 8

(d) Talc : 1

134. Match List-I with List-II and select the correct answer using the codes given below the Lists:

List-I	**List-II**
(Plagioclase feldspar)	(Mole% Anorthite)
(A) Oligoclase	1. An30- An50
(B) Labradorite	2. An10- An30
(C) Andesine	3. An50- An70
(D) Bytownite	4. An90- An100
	5. An70- An90

Code:

	A	B	C	D
(a)	5	1	3	4
(b)	2	1	4	5
(c)	5	3	2	4
(d)	2	3	1	5

135. Which one of the following chemical composition denotes the amphibole group of minerals?

(a) $(XY)Z_2O_5$ (b) $(XY)_{7\text{-}8}(Z_4O_{11})(OH)_2$

(c) $X_3Y_2Z_3O_{12}$ (d) $(XY)_{7\text{-}8}(Z_4O_{11})$

136. Which Garnet contains Chromium?

(a) Grossularite (b) Pyrope

(c) Uvarovite (d) Almandite

137. Which of the following mica mineral contains appreciable amounts of Li content?

(a) Biotite (b) Paragonite

(c) Muscovite (d) Lepidolite

138. Match List-I with List-II and select the correct answer using the codes given below the Lists:

List-I	**List-II**
(Mineral name)	(Chemical composition)
(A) Magnetite	1. Al_2SiO_5
(B) Dolomite	2. Al_2O_3
(C) Sillimanite	3. Fe_3O_4
(D) Corundum	4. $CaMg(CO_3)_2$

Code:

	A	B	C	D
(a)	2	1	4	3
(b)	3	1	4	2
(c)	2	4	1	3
(d)	3	4	1	2

139. The optic axis interference figure (uniaxial crystal) has:

(1) Radial lines indicate vibration direction of 'E' Ray.

(2) Show isochromatic curves.

(3) Tangential lines indicate vibration directions of 'O' Ray.

Select the correct answer from the code given below:

(a) 1,2 and 3
(b) 1 only
(c) 1 and 2 only
(d) 2 and 3 only

140. What is the correct term to denote the separate SiO_4 groups, in which the tetrahedral are not linked directly to each other but are bound together by intervening cations?

(a) Nesosilicates
(b) Sorosilicates
(c) Cyclosilicates
(d) Inosilicates

141. Which one of the following is not a phosphate mineral?

(a) Francolite
(b) Dahllite
(c) Collophane
(d) Phosphorite

142. Which of the following axis system is being satisfied by triclinic crystal system?

(a) $a = b = c, \alpha = \beta = \gamma = 90^o$

(b) $a = b = c, \alpha = \beta = \gamma = 90^o$

(c) $a \neq b \neq c, \alpha \neq \beta \neq \gamma \neq 90^o$

(d) $a = b = c, \alpha \neq \beta = \gamma = 90^o$

143. Which one of the following is most symmetrical?

(a) Simple cubic cell
(b) Hexagonal
(c) Triclinic
(d) Tetragonal

144. Which one of the following is least symmetrical?

(a) Tetragonal
(b) Simple cubic
(c) Triclinic
(d) Monoclinic

145. The incorrectly matched pair among the following

(a) Nesosilicates forsterite

(b) Cyclosilicates talc

(c) Phyllosilicates muscovite

(d) Tectosilicates orthoclase

146. Which of the following is not a mineral?

(a) Olivine (b) Talc

(c) Calcite (d) Quartzite

147. Which is not a polymorph of Al_2SiO_5?

(a) Kyanite (b) Calcite

(c) Silimanite (d) Andalusite

148. Which mineral shows double hardness?

(a) Magnesite (b) Quartz

(c) Feldspar (d) Kyanite

149. Hardness of fluorite is

(a) 3 (b) 4

(c) 7 (d) 5

150. Augite crystallized in which crystal system?

(a) Isometric system (b) Monoclinic

(c) Triclinic (d) Orthorhombic

151. The maximum number of axes of symmetry found in crystal system

(a) 13 (b) 10

(c) 22 (d) 9

152. Match the crystal system with correct number of

Group I (crystal system)	Group II (Classes)
P. Isometric	1. 3
Q. Orthorhombic	2. 12
R. Hexagonal	3. 7
S. Tetragonal	4. 5

(a) P-1, Q-2, R-3, S-4

(b) P-4, Q-1, R-3, S-2

(c) P-3, Q-4, R-2, S-1

(d) P-4, Q-1, R-2, S-3

153. Total number of classes in crystal system is

(a) 22 (b) 6

(c) 32 (d) 13

154. .... is an imaginary plane which passes through the center of crystal and divides it into two equal parts where one part is the mirror images of the other.

(a) Axial plane (b) Reference plane

(c) Plane of symmetry (d) Twin plane

155. Which one is not a physical property of mineral?

(a) Hardness (b) Streak

(c) From (d) Extinction

156. Which crystal is the non-pleochroic system?

(a) Hexagonal (b) Triclinic

(c) Cubic (d) Tetragonal

157. Biaxial minerals are not belong to ....crystal system?

(a) Triclinic (b) Tetragonal

(c) Monoclinic (d) Orthorhombic

158. Which mineral shows the double refraction?

(a) Calcite (b) Quartz

(c) Feldspar (d) Galena

159. Streak of hematite is

(a) Black colour (b) Yellow

(c) Cherry red (d) None of these

160. Fibrous form is characteristic for which mineral?

(a) Chrysotile (b) Hornblende

(c) Actinlite (d) Tremolite

161. The streak of a mineral is

(a) The colour of its powder

(b) The cleavage in monochromatic light

(c) The braking due to pressure

(d) Hardness in different directions

162. In which mineral there is no cleavage?

(a) Garnet (b) Spinel

(c) Quartz (d) All of these

163. Aquamarine is a variety of

(a) Corundum (b) Microcline

(c) Tourmaline (d) Beryl

164. Cyclic twinning is commonly seen in

(a) Aragonite (b) Staurolite

(c) Plagioclase (d) Gypsum

165. Staurolite is characteristic of which rock type?

(a) Igneous (b) Sedimentary

(c) Metamorphic (d) None of these

166. Which mineral has no silicate structure?

(a) Quartz (b) Calcite

(c) Augite (d) Muscovite

167. Which one of the following is an orthopyroxene?

(a) Enstatite (b) Aegirine

(c) Tremolite (d) Diopside

168. Of the following the first mineral to form, according to Bowen's reaction series.

(a) Hornblende (b) Albite

(c) Quartz (d) Olivine

169. Which one of the following contains manganese?

(a) Spessartite (b) Almandine

(c) Andradite (d) Grossularite

170. Which of the following is an incongruently melting compound?

(a) Orthoclase (b) Olivine

(c) Albite (d) Tourmaline

171. Which of the following is not a variety of gypsum?

(a) Alabaster (b) Selenite

(c) Satin spar (d) Iceland spar

172. Which one of the following is high temperature potassium feldspar?

(a) Orthoclase (b) Albite

(c) Sanidine (d) Anorthite

173. Orthoclase shows which type of twinning?

(a) Carlsbad (b) Baveno

(c) Manebach (d) All of these

174. The mineral fuchsite belongs to which group?

(a) Mica (b) Amphibole

(c) Pyroxene (d) Garnet

175. Which of the following feldspars does not contain potassium?

(a) Oligoclase (b) Sandine

(c) Orthoclase (d) Amazonstone

176. Calcium is not present in which one of the following minerals?

(a) Apatite (b) Anorthite

(c) Acmite (d) Actinolite

177. Which one of the following is radioactive?

(a) Microcline (b) Talc

(c) Monazite (d) Asbestos

178. Which one of the following is the softest mineral?

(a) Chlorapatite (b) Satin spar

(c) Bort (d) Iceland spar

179. The mineral wollastonite crystallizes in which crystal system?

(a) Orthorhombic (b) Monoclinic

(c) Triclinic (d) Hexagonal

180. Which of the following is used in crystallography?

(a) Weiss symbol (b) Hermann Maugin notation

(c) Miller indices (d) All of the above

181. Except for centre of symmetry the other axial elements are absent in

(a) Tourmaline (b) Microcline

(c) Gypsum (d) Orthoclase

182. What will be the miller symbol of face whose intercepts on the axes are a: 5b: 4c?

(a) 45 (b) 045

(c) 045 (d) 154

183. The general symbol for a pyritohedron is

(a) (*h00*) (b) (*hk0*)

(c) (*hkl*) (d) (*001*)

184. A five-fold axis of symmetry is present in

(a) Quartz (b) Riebeckite

(c) Tourmaline (d) None of these

185. Of the feldspars, Orthoclase, Microcline, Anorthoclase, Albite, Andesine, Bytownite, Anorthite, Hyalophane and Celsian, how many are monoclinic?

(a) Two (b) Three

(c) Four (d) Five

186. In how many crystal systems the c-axis makes an angle with the b-axis?

(a) One (b) Two

(c) Three (d) Four

187. The angle between the c-axis and c-axis is called

(a) α (b) β

(c) γ (d) δ

188. For which crystal system the miller symbol for the basal pinacoid is not 001?

(a) Tetragonal (b) Hexagonal

(c) Monoclinic (d) Triclinic

189. Of the thirty-two classes of symmetry, which of the following system has three classes?

(a) Tetragonal (b) Hexagonal
(c) Monoclinic (d) Triclinic

190. The number of crystallographic axes in uniaxial minerals is

(a) Three (b) Three of four
(c) Four (d) One

191. 2V is the angle between the

(a) a and c crystallographic axes (b) Biaxial optic axes
(c) Fastest and slowest direction (d) None of the above

192. A mineral 'X' is invisible in Canada Balsam. The refractive index of 'X' is

(a) 1.54 (b) 1.44
(c) 1.64 (d) 1.00

193. The highest relief will be shown by which one of the following minerals?

(a) 'A' RI = 1.42 (b) 'B' RI = 1.44
(c) 'C' RI = 1.59 (d) 'D' RI = 1.62

194. A mineral is in extinction position when its vibration plane is parallel to

(a) Vibration direction of polarizer
(b) Vibration direction of analyser
(c) Hexagonal system
(d) Orthorhombic system

195. A mineral has three values for its refractive indices. The mineral crystallizes in

(a) Cubic system (b) Tetragonal system
(c) Hexagonal system (d) Orthorhombic system

196. In uniaxial minerals, there is no double refraction along which crystallographic axis?

(a) a-axis (b) b-axis
(c) c-axis (d) All of these

197. The optic axes in biaxial minerals always lie in

(a) X-Y plane (b) X-Z plane
(c) Y-Z plane (d) Any of the above combinations

198. Which mineral shows one set of cleavage in thin sections?

(a) Muscovite (b) Calcite
(c) Wollastonite (d) Microcline

199. Which mineral commonly shows euhedral shape?

(a) Quartz (b) Zircon
(c) Microcline (d) Muscovite

200. Inclined extinctions are shown by which of the following mineral sets?

(a) Andesine, Hornblende, Kyanite
(b) Augite, Muscovite, Beryl
(c) Biotite, Tourmaline, Quartz
(d) Gernet, Microcline, Hypersthene

201. Which mineral is isotropic under the microscope?

(a) Almandine (b) Biotite
(c) Tourmaline (d) All of these

202. Isochromatic lines are seen as a part of the interference figure in

(a) Tetragonal minerals (b) Hexagonal minerals
(c) Monoclinic minerals (d) All of the above

203. Calcite is

(a) Uniaxial positive (b) Uniaxial negative
(c) Biaxial positive (d) Biaxial negative

204. Inclined extinction is shown by

(a) Calcite (b) Basal section of augite
(c) Basal section of hornblende (d) None of the above

205. The simplest of all the silicate structures is

(a) Orthosilicates (b) Sorosilicates
(c) Metasilicates (d) Inosilicates

206. Which one is not a potash feldspar?

(a) Orthoclase (b) Oligoclase

(c) Sanidine (d) Microcline

207. The main difference between graphite and diamond is

(a) Composition (b) Crystal structure

(c) Density (d) Colour

208. Which of the following minerals is having the chemical formula (Fe, Mg) SiO_3

(a) Andalusite (b) Topaz

(c) Hypersthene (d) Barite

209. Which one of the following is not an example of triclinic system?

(a) Kyanite (b) Albite

(c) Andalusite (d) Microcline

210. Mineral diamond crystallizes in

(a) Orthorhombic system (b) Tetragonal system

(c) Cubic system (d) Monoclinic system

211. Which of the following systems has all closed forms?

(a) Triclinic (b) Cubic

(c) Trigonal (d) Monoclinic

212. Which of the following has 3 axes of 4 fold symmetry?

(a) Baryte (b) Gypsum

(c) Galena (d) Rutile

213. Which one is isotropic mineral?

(a) Quartz (b) Garnet

(c) Tourmaline (d) Gypsum

214. In an uniaxial positive mineral the velocity of ordinary ray is

(a) Greater than that of extraordinary ray

(b) Equal to that of extraordinary ray

(c) Less than that of extraordinary ray

(d) Neither less nor equal to that of extraordinary ray

215. The chemical composition of jadeite is

(a) $MgSiO_3$ (b) $NaFe(SiO_3)_2$

(c) $NaAl(SiO_3)_2$ (d) $LiAl\ (SiO_3)_2$

216. The hardness of orthoclase on Moh's scale is

(a) 7 (b) 6

(c) 5 (d) 8

217. The end members of orthorhombic pyroxene are

(a) Enstatite-diopside (b) Enstatite hypresthene

(c) Diopside augite (d) Augite jaedite

218. Fibrous variety of quartz is

(a) Flint (b) Chalcedony

(c) Chert (d) Amethyst

219. Which of the following is not a magnetic mineral?

(a) Pyrrhotite (b) Hematite

(c) Orthoclase (d) Magnetite

220. Which one is a light mica?

(a) Pholgopite (b) Biotite

(c) Paragonite (d) Zinwaldite

221. Olivine belongs to

(a) Neosilicate (b) Insocilicate

(c) Sorosilicate (d) Cyclosilicate

222. What is mineral 'wool'?

(a) Albite (b) Andalusite

(c) Asbestos (d) Kyanite

223. The native mineral having hackly fracture is

(a) Sulphur (b) Copper

(c) Gold (d) Borax

224. Which of the following properties is not observed under ordinary light?

(a) Colour (b) Inclusions

(c) Pleochroism (d) Refractive Index (RI)

225. Augite shows

(a) 1st order interference colours

(b) 2nd order interference colours

(c) 3rd order interference colours

(d) 4th order interference colours

226. The indicatrix of a uniaxial positive crystal is a/a

(a) Sphere (b) Ellipsoid

(c) Oblate (d) Prolate spheroid

227. Vertical section of uniaxial mineral shows

(a) Dichroism (b) Pleochroism

(c) Isotropism (d) Equal illumination

228. Refringence is the ability to produce

(a) Reflection (b) Extinction

(c) Refraction (d) Interference

229. The most fundamental and common form of pyroxene is

(a) Pyramid (b) Prism

(c) Basal pinacoid (d) Side pinacoid

230. Find the odd one out

(a) Biotite (b) Phlogopite

(c) Enstatite (d) Muscovite

231. The silicate class has ……… subclasses.

(a) 4 (b) 6

(c) 8 (d) 10

232. Leucoxene is veriety of

(a) Tourmaline (b) Talc

(c) Sphene (d) Topaz

233. Hardness of Talc is.

(a) 1 (b) 2

(c) 5 (d) 6

234. The highest degree of symmetry is shown by and the lowest degree of symmetry is shown by _____respectively

(a) Cubic system and Triclinic system

(b) Cubic system and Monoclinic system

(c) Hexagonal system and Monoclinic system

(d) Tetragonal system and Monoclinic system

235. Tetragonal system is characterised by

(a) Plane of symmetry (b) Axis of symmetry.

(c) Centre of symmetry (d) All the above.

236. Where would you expect to find the largest crystals in a lava flow?

(a) Near the top surface of the flow

(b) In the center of the flow

(c) Near the bottom of the flow

(d) The crystals would have the same grain size throughout the flow

237. According to Bowen's reaction series, which of the following pairs of phases are likely to be incompatible?

(a) Quartz and alkali feldspar (b) Ca-Plagioclase and olivine

(c) Quartz and olivine (d) Na-plagioclase and amphibole

238. During crystallization of a melt plagioclase become richer in

(a) Potassium (b) Sodium

(c) Calcium (d) Krypton

239. Crystal face with miller indices (111) is known as

(a) Parametral face (b) Solid face

(c) Inclined face (d) Unit face

240. The chemical composition of Jadeite is

(a) $NaAl(SiO_3)_2$ (b) $MgSiO_3$

(c) $NaFe(SiO_3)_2$ (d) $LiAl(SiO_3)_2$

241. Line perpendicular to a circular section of an indicatrix is

(a) Twin axis (b) Optic axis

(c) Rotational axis (d) Axis of symmetry

242. The mineral Augite shows

(a) 1st order interference colours

(b) 2nd order interference colours

(c) 3rd order interference colours

(d) 4th order interference colours

243. Which one is a non-pleochroic mineral?

(a) Biotite (b) Olivine

(c) Garnet (d) Tourmaline

244. The end members of orthorhombic pyroxene are

(a) Enstatite-Diopside (b) Diopside-Augite

(c) Enstatite-Hypersthene (d) Augite-Jadeite

245. High-temperature polymorph of silica is

(a) Cristobalite (b) Flint

(c) Chert (d) Amethyst

246. The chemical composition of Orthoclase is

(a) $K\,AlSi_2O_4$ (b) $K\,Al\,Si_3O_6$

(c) $K\,AlSi_4O_8$ (d) $KAl\,Si_3O_8$

247. Chromite is a member mineral of

(a) Epidote group (b) Spinel group

(c) Olivine group (d) Mellilite group

248. Olivine belongs to

(a) Nesosilicate (b) Inosilicate

(c) Sorosilicate (d) Cyclosilicate

249. A texture in which phenocrysts are embedded in fine grained ground mass is

(a) Porphyritic (b) Perthite

(c) Graphic texture (d) Seriate texture

250. Which of the following mineral exhibits the property of magnetism?

(a) Biotite (b) Orthoclase

(c) Albite (d) Pyrrhotite

251. Which one of the following is an ore mineral of silver?

(a) Arsenopyrite (b) Argentite
(c) Cassiterite (d) Cerussite

252. Which one of the following is a chromium-bearing garnet?

(a) Grossular (b) Uvarovite
(c) Spessartine (d) Pyrope

253. Which one of the following minerals shows piezoelectric property?

(a) Garnet (b) Apatite
(c) Quartz (d) Orthoclase- feldspar

254. In isometric system, a six faced solid in which each face cuts one axis and is parallel to the other two axes is known as:

(a) Hexahedron (b) Octahedron
(c) Rhombic dodecahedron (d) Tetrahexahedron

255. An open form comprising two horizontal faces in normal class of Tetragonal system is termed as:

(a) Pyramid (b) Dome
(c) Basal Pinacoid (d) Prism

256. Which one of the following is the correct composition of Troctolite?

(a) Olivine +Plagioclase
(b) Olivine +Clinopyroxene
(c) Clinopyroxene +Plagioclase
(d) Clinopyroxene +Orthopyroxene

257. The degrees of freedom at peritectic point in binary peritectic system will be:

(a) Three (b) Two
(c) One (d) Zero

258. Mercury is the path finder element for:

(a) Copper deposits (b) Lead- Zinc- Silver deposits
(c) Chromite deposits (d) Uranium deposits

259. Which one of the following physical properties is NOT related to mineral asbestos?

(a) Readily separable into fine filaments

(b) High tensile strength

(c) Enough flexibility

(d) High degree of combustibility

260. Which among the following mineral displays twinkling?

(a) Augite

(b) Calcite

(c) Gypsum

(d) Talc

261. Which one of the following mineral exhibits admantine luster?

(a) Nepheline (b) Quartz

(c) Calcite (d) Diamond

262. Quartz has:

(a) Even fracture (b) Uneven fracture

(c) Hackly fracture (d) Conchoidal fracture

263. Olivine belongs to

(a) Sorosilicate (b) Nesosilicate

(c) Inosilicate (d) Cyclosilicate

264. Mineral diamond crystallizes in:

(a) Monoclinic system (b) Tetragonal system

(c) Orthorhombic system (d) Cubic system

265. Pericline twinning is present in which minerals?

(a) Calcite (b) Staurolite

(c) Plagioclase (d) None of these

266. The triclinic system is characterized by

(a) No plane or axes of symmetry

(b) 4 axes of 3 fold symmetry

(c) 1 axies of 3 fold symmetry
(d) Only one plane of symmetry

267. Trigonal symmetry is characterized by

(a) 4 plane, 4 axes and no center
(b) No planes, 4 axes and no center
(c) No plane, 4 axes, and a center
(d) None of these

268. The symmetry of crystal depends upon

(a) Shape of the faces
(b) Size of the faces
(c) Distribution of angular elements
(d) Crystallographic notation

269. The highest degree of symmetry is down by the

(a) Triclinic system (b) Isometric system
(c) Hexagonal system (d) Tetragonal system

270. The bravis lattice of sodium chloride (NaCl) structure is

(a) Simple cube (b) Body centred cube unit
(c) Base centred cube (d) Face centred cube

271. How many atoms per cell are present in a body centred lattice?

(a) Three atoms per unit cell
(b) One atom per unit cell
(c) Two atoms per unit cell
(d) Four atoms per unit cell

ANSWER SHEET

1. (b)	2. (c)	3. (b)	4. (a)	5. (c)
6. (d)	7. (b)	8. (d)	9. (d)	10. (c)
11. (d)	12. (a)	13. (d)	14. (a)	15. (b)
16. (c)	17. (a)	18. (d)	19. (b)	20. (d)
21. (a)	22. (a)	23. (b)	24. (c)	25. (a)
26. (d)	27. (a)	28. (d)	29. (a)	30. (a)
31. (b)	32. (b)	33. (b)	34. (c)	35. (a)
36. (c)	37. (a)	38. (a)	39. (c)	40. (b)
41. (a)	42. (b)	43. (b)	44. (b)	45. (d)
46. (b)	47. (b)	48. (b)	49. (c)	50. (b)
51. (a)	52. (a)	53. (b)	54. (a)	55. (c)
56. (a)	57. (a)	58. (a)	59. (a)	60. (b)
61. (d)	62. (c)	63. (b)	64. (a)	65. (a)
66. (d)	67. (b)	68. (b)	69. (c)	70. (c)
71. (a)	72. (c)	73. (d)	74. (c)	75. (a)
76. (c)	77. (b)	78. (d)	79. (c)	80. (c)
81. (d)	82. (a)	83. (b)	84. (d)	85. (d)
86. (d)	87. (c)	88. (b)	89. (d)	90. (b)
91. (c)	92. (c)	93. (d)	94. (b)	95. (a)
96. (b)	97. (d)	98. (d)	99. (a)	100. (c)
101. (d)	102. (d)	103. (d)	104. (b)	105. (d)
106. (c)	107. (d)	108. (d)	109. (c)	110. (d)
111. (c)	112. (c)	113. (b)	114. (c)	115. (c)
116. (d)	117. (c)	118. (c)	119. (c)	120. (c)
121. (c)	122. (a)	123. (b)	124. (d)	125. (a)
126. (d)	127. (b)	128. (d)	129. (d)	130. (a)
131. (c)	132. (c)	133. (c)	134. (d)	135. (b)
136. (c)	137. (d)	138. (d)	139. (a)	140. (a)
141. (a)	142. (c)	143. (a)	144. (c)	145. (b)
146. (d)	147. (b)	148. (d)	149. (b)	150. (b)
151. (a)	152. (d)	153. (c)	154. (c)	155. (d)
156. (c)	157. (b)	158. (a)	159. (c)	160. (a)
161. (a)	162. (d)	163. (d)	164. (a)	165. (c)
166. (b)	167. (a)	168. (d)	169. (a)	170. (a)
171. (d)	172. (c)	173. (d)	174. (a)	175. (a)
176. (c)	177. (c)	178. (b)	179. (c)	180. (d)
181. (b)	182. (b)	183. (b)	184. (d)	185. (b)

186. (a)	187. (b)	188. (b)	189. (c)	190. (b)
191. (b)	192. (a)	193. (a)	194. (c)	195. (d)
196. (c)	197. (b)	198. (a)	199. (b)	200. (c)
201. (a)	202. (d)	203. (b)	204. (d)	205. (a)
206. (b)	207. (b)	208. (c)	209. (c)	210. (c)
211. (b)	212. (c)	213. (b)	214. (a)	215. (c)
216. (b)	217. (b)	218. (b)	219. (c)	220. (c)
221. (a)	222. (c)	223. (b)	224. (c)	225. (b)
226. (d)	227. (a)	228. (c)	229. (b)	230. (c)
231. (b)	232. (c)	233. (a)	234. (a)	235. (b)
236. (b)	237. (c)	238. (b)	239. (d)	240. (a)
241. (b)	242. (b)	243. (c)	244. (c)	245. (a)
246. (d)	247. (b)	248. (a)	249. (a)	250. (d)
251. (b)	252. (b)	253. (c)	254. (a)	255. (c)
256. (a)	257. (d)	258. (b)	259. (d)	260. (b)
261. (d)	262. (d)	263. (b)	264. (d)	265. (c)
266. (a)	267. (b)	268. (c)	269. (b)	270. (d)
271 (c)				

CHAPTER - 4

PETROLOGY

(A) IGNEOUS

(B) SEDIMENTARY

(C) METAMORPHIC

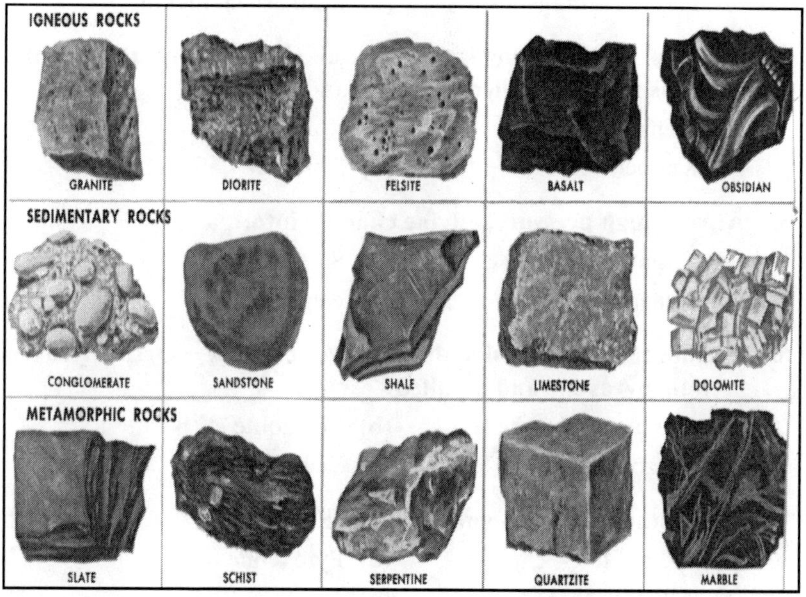

PETROLOGY

(A) IGNEOUS PETROLOGY

1. Which of the following rocks represents maximum amount of radioactive heat flow?

(a) Dolerite (b) Basalt
(c) Granite (d) Syenite

2. In discontinuous reaction series of Brown, the Na, K/Al value attains to 1 in case of crystallization of which of the following minerals?

(a) Amphibole (b) Forsterite
(c) Pyroxene (d) Biotite

3. Reabsorbed texture in forsterite is due to:

(a) Magma mixing (b) Magma mingling
(c) Partial melting (d) Incongruent melting

4. Which of the following is the mafic alkaline porphyritic dyke that contains mica, hornblende, augite and olivine?

(a) Lamprophyre (b) Komatite
(c) Kimberlite (d) Dolerite

5. At very high pressure, olivine changes into:

(a) Enstatite (b) Spinel
(c) Hypersthene (d) Augite

6. Peridotite is composed of 40% - 90% of olivine + Ortho pyroxene + clinic pyroxene and is called:

(a) Olivine websterite (b) Lherzolite
(c) Harzburgite (d) Websterite

7. Hypersthene bearing gabbro is called:

(a) Troctolite (b) Pyroxenite
(c) Alkali gabbro (d) Norite

8. Lipartite is a synonym for which of the following rocks?

(a) Dolerite (b) Dunite
(c) Rhyolite (d) Syenite

9. Match items from List-I and List-II and select the correct answer using the code given below:

List-I		List-II	
(a)	Oligoclase	(I)	Au30- Au50
(b)	Labradorite	(II)	Au10- Au30
(c)	Andesine	(III)	Au50- Au70
(d)	Bytownite	(IV)	Au90- Au100
		(V)	Au70- Au90

Code:

	A	B	C	D
(1)	II	I	III	IV
(2)	II	I	IV	V
(3)	V	III	II	IV
(4)	II	III	I	V

10. Which of the following condition is true for Norite?

(a) CPX is absent
(b) OPX < CPX
(c) OPX = CPX
(d) OPX is absent

11. What is tektite?

(a) Glassy meteorite
(b) Sedimentary rock
(c) Metamorphic rock
(d) Tectonically deformed rock

12. The temperature at which a magnetic body completely loses its magnetism is called -

(a) Curie temperature
(b) Blocking temperature
(c) Homogenization temperature
(d) Eutectic temperature

13. When K- feldspar is rimmed by sodic plagioclase the texture is known as ...

(a) Rapa kivi
(b) Graphic
(c) Ophitic
(d) Orbicular

14. Which of the following is correct for S-type granite?

(a) They are peraluminous
(b) They have high Na_2O / K_2O ratio
(c) They always contain hornblende
(d) They are formed from a metabasic source

15. Sub ophitic texture is characteristic of

(a) Sand stone
(b) Lime stone
(c) Dolerite
(d) Granite

16. In which of the following rocks Mg should be <80%

(a) Diorite
(b) Grano-diorite
(c) Kanotite
(d) Basalt

17. Malani Igneous suite is mainly composed of

(a) Lava flows of basalt
(b) Acidic volcanic rocks
(c) Ultrabasic rocks
(d) Granites

18. The most common mineral found in igneous rock is:

(a) Feldspar
(b) Olivine
(c) Muscovite
(d) Fluorite

19. An igneous rock which is commonly black, has a glassy texture, and has a conchoidal fracture is which of the following?

(a) Pelite
(b) Rhyolite
(c) Obsidian
(d) None of the above

20. Which of the following is the most important chemical active fluid involved in the formation of rocks?

(a) Hydrochloric acid
(b) Sulfuric acid
(c) Water
(d) Methane

21. A type of volcano composed of mainly pyroclastic is termed:

(a) Cinder cone
(b) Shield cone
(c) Composite cone
(d) None of these

22. Which of the following pairs contains one igneous and one sedimentary rock?

(a) Shale and marble
(b) Sandstone and quartzite
(c) Granite and limestone
(d) Obsidian and gneiss

23. Which of these pairs of minerals are always found in granites?

(a) Muscovite and calcite (b) Quartz and orthoclase

(c) Hornblende and talc (d) Augite and magnetite

24. Which of the following is an igneous rock?

(a) Granite (b) Shale

(c) Coal (d) Slate

25. Igneous rocks that form from magma are known as:

(a) Minerals (b) Granite

(c) Intrusive rocks (d) Gneiss

26. What are tektites?

(a) Glassy meteorites (b) Sedimentary rocks

(c) Igneous rocks (d) Tectonically deformed rocks

27. When the mineral Augite encloses thin laths of plagioclase, the texture is named as-

(a) Trachytic texture (b) Ophitic texture

(c) Anatectic texture (d) Felsitic texture

28. Nepheline - Feldspar bearing Garbbroic rock is known as -

(a) Norite (b) Charnockite

(c) Dunite (d) Essexite

29. Dominant ortho-pyroxene bearing gabbro is named as-

(a) Pyroxenite (b) Norite

(c) Latite (d) Phonolite

30. Ortho-pyroxene bearing granite is known as-

(a) Charnockite (b) Dolerite

(c) Synite (d) Essexite

31. The Jalore granite is considered equivalent to

(a) Banded Gneissic Complex

(b) Bearch Granite

(c) Mirpur Granite

(d) Darwal Granite

32. In Rajasthan Granulite rocks are present at which place -

(a) Siwana (b) Mokalsar

(c) Sandmata (d) Erinpura

33. The Mylonite develops due to -

(a) Thermal metamorphism (b) Ductile shearing

(c) Weathering effect (d) Chemical effect

34. A Dark Coloured porphyritic mafic to ultramafic Potassium rich hypabyssal rock is named as -

(a) Ignimbrite (b) Tachylite

(c) Lamprophyre (d) Felsite

35. Consider the following ultramafic rocks (Igneous):

1. Peridotite 2. Alkali-Peridotite

3. Dunite 4. Pyroxenite

What is the correct sequence of these in the order of decreasing MgO and increasing CaO?

(a) 3-2-1-4 (b) 4-1-2-3

(c) 3-1-2-4 (d) 4-2-1-3

36. Which one of the following textures characterizes the Eutectic point?

(a) Porphyritic (b) Ophitic

(c) Subophitic (d) Graphic

37. Which of the following rocks is noted for having mantle xenoliths in it?

(a) Carbonatite (b) Lamprophyre

(c) Kimberlite (d) Komatite

38. F (Norm) values of Birth are calculated to determine which one of the following?

(a) Rate of crystallization of magma

(b) Depth of crystallization of magma

(c) Priority of crystallization of plagioclase or pyroxene from a basic magma

(d) Composition of plagioclase and pyroxene crystallizing from magma

39. Match List-I with List-II and select the correct answer using the codes given below the Lists:

List-I (Rock)		List-II (% Mineral)
(A) Websterite	1.	60% ol- 15% opx- 25% cpx
(B) Lherzolite	2.	60% ol- 10% opx- 30% cpx
(C) Harzburgite	3.	60% ol- 30% opx- 10% cpz
(D) Wehrlite	4.	10% ol- 50% opx- 40% cpx

Code:

	A	B	C	D
(a)	2	3	1	4
(b)	4	3	1	2
(c)	2	1	3	4
(d)	4	1	3	2

40. What is the curve above which the system is completely liquid?

(a) Solidus
(b) Liquidus
(c) Eutectic point
(d) Solid Solution

41. Which one of the following can be used as way-up-criteria?

(a) Pillow lava
(b) Blocky lava
(c) Ropy lava
(d) Pa-hoe-hoe-lava

42. Which one of the following represents an intrusion of the magmas of different composition through the same channel?

(a) Multiple intrusion
(b) Composite intrusion
(c) Forced intrusion
(d) Dyke

43. Who discussed the origin of Basalts in terms of the An-Di-Ab-Fo quaternary system?

(a) Osborn and Tait (1952)
(b) Yoder and Tilley (1962)
(c) Tuttle and Bowen (1958)
(d) Bowen and Schenirer (1935)

44. Consider the following:

1. Perfect gradation between the country rocks and the granites.
2. Chilled margins of the Granite.
3. Presence of xenoliths of the earlier country rocks in the granites.
4. Granophyric or graphic texture
5. Relict bedding in Granitic rocks.

Which of these are the evidences in favour of the magmatic origin of granites?

(a) 1, 2 and 3 (b) 1, 2 and 4

(c) 2, 3 and 4 (d) 3, 4 and 5

45. Which of the following petrographic descriptions fits the term Quartz arenite?

(a) 95% quartz and matrix content greater than 15%

(b) 95% quartz and matrix content less than 15%

(c) 85% quartz and feldspar content greater than 10%

(d) 85% quartz and rock fragment content greater than 10%

46. Which one of the following retrograde metamorphic pairs is not correctly matched?

(Reactant) (Product)

(a) Diopside: Tremolite/ Actinolite

(b) Grossular Garnet: Biotite, Chlorite or Serpentine

(c) Kyanite: Sericite/ Muscovite + Quartz

(d) Plagioclase: Zoisite/ Clinizoisite + Albite

47. Which of the following mineral assemblages indicate granulite facies?

(a) Quartz - Albite - Epidote - Muscovite

(b) Quartz - Orthoclase - Sillimanite - Almandine

(c) Calcite - Quartz - Epidote - Tremolite

(d) Quartz - Oligoclase - Muscovite - Andalusite - Cordierite

48. What is the degree of freedom at the eutectic point in a binary System?

(a) 1 (b) 0
(c) 3 (d) 2

49. Which one of the following processes may result in development of disjunctive cleavage?

(a) Crystallization (b) Dislocation Creep
(c) Pressure Solution (d) Grain boundary migration

50. Spinifex texture is found in which rock?

(a) Lamprophyre (b) Komatite
(c) Tholeiite (d) Rhyolite

51. Dolerite shows which texture?

(a) Ophitic (b) Panidiomorphic
(c) Allotrimorphic (d) Seriate

52. Eutectic crystallization of quartz and K-feldspar give rise to:

(a) Hypidiomorphic texture (b) Porophyritic texture
(c) Graphic texture (d) All of the above

53. In the IUGS classification, the field of diorite falls near which end of the triangle?

(a) Quartz (Q) (b) Alkali feldspar (A)
(c) Plagioclase (P) (d) None of the above

54. The periclase silica phase diagram exhibits

(a) Eutectic points (b) Peritectic point
(c) Liquid immiscibility (d) All of the above

55. As per the Bowen's reaction series the correct order of crystallization is shown by:

(a) Albite-Anorthite-Labradorite
(b) Anorthite-Bytownite-Labradorite
(c) Albite-Oligoclase-Andesine
(d) Anorthite-Labradorite-Bytownite

56. Which one is a type of meteorite?

(a) Anthracite (b) Anticilite
(c) Aubrite (d) Adulite

57. Sideromelane is a synonym for:

(a) Tachylyte (b) Basinite
(c) Boninite (d) Trachyte

58. A komatiite should contain

(a) CaO>16% (b) $Na_2O+K_2O>22\%$
(c) $SiO_2>54\%$ (d) MgO>18%

59. Filter pressing is associated with

(a) Magmatic differentiation (b) Texture
(c) Volatiles (d) Viscosity

60. Pillow structure is shown by

(a) Pillow (b) Rhyolites
(c) Lamprophyres (d) Basaltic lava

61. Ankaramite and picrite are types of

(a) Basalts (b) Granite
(c) Fossils (d) Lamprophyres

62. The condensed phase rule is given by the formula

(a) $P + C = F - 1$ (b) $1 - P = F - C$
(c) $F + C = P - 2$ (d) $P + F = C + 2$

63. Viscosity of a magma increases with

(a) Increase in Mg (b) Increase in Al
(c) Increase in H_2O (d) All of the above

64. A magmatic association with felsic and mafic members predominating is termed as

(a) Binomial (b) Bi-magmatic
(c) Bimodal (d) Bi-petrotectonic

65. Which of the following one is not a meteorite?

(a) Chondrite (b) Impactite
(c) Achondrite (d) Siderolite

66. Dacite is the volcanic equivalent of

(a) Granite (b) Syenite

(c) Diorite (d) Granodiorite

67. In a phase diagram, the phase rule at an invariant point is defined by

(a) F = 0 (b) F = 1

(c) F = 2 (d) F = 3

68. Which of the following are lithophile elements?

(a) Ag, Zn, Pb (b) Li, Zr, V

(c) Ru, Os, Au (d) S, Se, Te

69. Elements of B-subgroups (in periodic table) with 18 electrons in outermost shell are:

(a) Atmophile (b) Lithophile

(c) Chalcphile (d) Siderophile

70. Rubidium strontium Datin is useful for:

(a) Relatively young sediments

(b) Relatively old geological material

(c) A hundred to thousand-year-old rocks

(d) Non-geological material

71. Element of which atomic number is absent in the earth?

(a) 41 (b) 42

(c) 43 (d) 44

72. The weight percent oxygen in the earth crust is about:

(a) 42 (b) 46

(c) 49 (d) 53

73. Fine grained igneous rocks are

(a) Intrusive body (b) Plutonic

(c) Extrusive body (d) Sedimentary body

74. Hard, massive, compact with interlocking grains and absence of Fossic and bedding is the essential feature of:

(a) Igneous rocks (b) Sedimentary rocks

(c) Metamorphic (d) Limestone

75. Volcanic equivalent of granite is

(a) Basalt (b) Gabbro

(c) Rhyolite (d) Diorite

76. Gabbro is ....type of igneous rock.

(a) Volcanic (b) Plutonic

(c) Hypabyssal (d) Metasedimentary

77. Which one is not concordant body?

(a) Laccoliths (b) Lopoliths

(c) Phacoliths (d) Batholiths

78. Which mineral crystallized last from the magma?

(a) Olivine (b) Quartz

(c) Hornblende (d) Labradorite

79. The 'as structure' is also known as

(a) Pahoehoe (b) Ropy lava

(c) Blocky lava (d) Flow structure

80. The occurrence of foreign materials within igneous rocks are formed a structure that is

(a) Xenolithic structure (b) Myrmekite structure

(c) Sheet structure (d) Amygdaloidal structure

81. When the igneous rocks made of only glassy materials known as

(a) Holocrystalline (b) Hemi crystalline

(c) Holohyaline (d) Semihyaline

82. Perlitic cracks in crystal represent the conversion from

(a) Basic magma (b) Acidic magma

(c) Glassy condition (d) Plutonic condition

83. Eutectic point is a

(a) Univariant (b) Invariant

(c) Di-variant (d) Tri-variant

84. Silica content in basic igneous rock is

(a) 44 to 55% (b) More than 65%

(c) 55 to 65% (d) More than 80%

85. Gem quality olivine is termed as

(a) Norite (b) Satin-spar

(c) Peridot (d) Ruby

86. Intergrowth of quartz and feldspar resulting

(a) Perthite (b) Orbicular structure

(c) Graphic texture (d) Myrmekite structure

87. Correctly match the following:

Column-A	**Column-B**
Q- Laths of plagioclase are totally surrounded in a coarse grained matrix of pyroxene crystals	1. Spherulitic texture
R- Smaller grains of one mineral are completely enclosed in large	2- Ophitic texture
S- Plagioclase grains show a preferred orientation due to flowage	3- Poikilitic texture
T- Spherical intergrowths of radiating quartz and feldspar replace glass	4- Trachytic texture

(a) P-2, Q-3, R-4, S-1 (b) P-1, Q-2, R-3, S-4

(c) P-2, Q-4, R-3, S-1 (d) P-1, Q-3, R-2, S-4

88. Which one is most stable mineral?

(a) Quartz (b) Feldspar

(c) Zircon (d) Olivine

89. Why basalt is finer grained than gabbro's?

(a) Basalt formed from quick cooling of magma

(b) Gabbro formed from quick cooling of magma

(c) Basalt has a mafic composition

(d) Gabbro has a mafic composition

90. Rocks are formed when magma

(a) Erodes (b) Crystallizes

(c) Undergoes radioactive decay (d) Weathers

91. Basic source of magnetism is

(a) Charged particles alone

(b) Magnetic domain

(c) Magnetic dipoles

(d) Movement of charged particles

92. Obsidian is a/an

(a) Metamorphic rock which contain high mafic mineral

(b) Igneous rock which contain high Fe and Mg

(c) Volcanic glass containing high silica

(d) Sedimentary rock formed by weathering of basaltic rock

93. Match each items and choose the correct option:

	Column-A		**Column-B**
P.	Felsic	1.	Magma that flows out Earth's surface
Q.	Kimberlitic	2.	Dark coloured rock, such as gabbro that is rich in Fe and Mg
R.	Lava	3.	Rock that is rich in silica
S.	Mafic	4.	Ultramafic rock that can contains diamonds

(a) P-1, Q-3, R-4, S-2 (b) P-3, Q-4, R-1, S-2

(c) P-2, Q-1, R-4, S-3 (d) P-4, Q-1, R-3, S-2

94. The gabbroic rock without pyroxenes containing mainly feldspars and olivine is

(a) Norite (b) Troctolite

(c) Dunite (d) Eucrite

95. The volcanic rocks containing the highest percentage of silica is

(a) Rhyolites (b) Trachytes

(c) Andesites (d) Basalts

96. Granophyres are hypabyssal equivalent of

(a) Basalt (b) Granite
(c) Gabbro (d) Diorite

97. An example of a discordant igneous intrusion is

(a) Chonolith (b) Laccolith
(c) Lopolith (d) Bysmalith

98. Pegmatite rock contains one of the following in abundance

(a) Tourmaline (b) Spinel
(c) Forsterite (d) Andesine

99. A typical monomineralic rock is

(a) Syenite (b) Granite
(c) Anorthosite (d) Dolerite

100. A texture in which phenocrysts are embedded in fine grained ground mass is

(a) Perthite (b) Porphyritic
(c) Graphic (d) Seriate texture

101. Orthophyric texture is a type of

(a) Intergrowth texture (b) Inequulgranular texture
(c) Equigranular texture (d) Intergranular texture

102. Peridotite is

(a) An amphibole (b) A pyroxene
(c) An acid igneous rock (d) An ultramafic rock

103. Blue-granite is also known as

(a) Monozonite (b) Diorite
(c) Larvikite (d) Nordmarkite

104. Lavas containing numerous gas cavities of irregular shape are

(a) Scoria (b) Pumice
(c) Amygdales (d) Ignimbrites

105. The variety of peridotite, in which olivine is altered to serpentine, is

(a) Pyroxenite (b) Kimberlite
(c) Dunite (d) Bronzitites

106. Which of the following pairs of intrusive and extrusive rocks have the same chemical composition?

(a) Granite and andesite (b) Diorite and basalt
(c) Gabbro and basalt (d) Gabbro and rhyolite

107. The gabbroic rock without pyroxenes containing mainly feldspars and Olivine is:

(a) Andesite (b) Dacite
(c) Troctolite (d) Basalt

108. Batholiths are usually associated with:

(a) Orogenic belt (b) Earthquake zone
(c) Island arcs (d) Fold and faults

109. Ophitic texture is commonly exhibited by:

(a) Lamprophyres (b) Dolerite
(c) Andesite (d) Trachyte

110. Lavas containing numerous gas cavities of irregular shape are:

(a) Pumice (b) Amygdales
(c) Scoria (d) Ignimbrites

111. Granophyre's are hypabyssal equivalent of

(a) Basalt (b) Granite
(c) Gabbro (d) Diorite

112. An example of a discordant igneous intrusion is

(a) Laccolith (b) Lopolith
(c) Bysmalith (d) Chonolith

113. Which of the following are the examples of layered mafic intrusion?

(1) Bushveld igneous complex
(2) Still water complex
(3) Skaergaard intrusion

Select the correct answer using the code given below:

(a) 1 and 2 only (b) 2 and 3 only

(c) 1 and 3 only (d) 1, 2 and 3

114. Mineralogical composition of pulaskite rock is:

(a) Olivine + augite + calc. plagioclase

(b) Antiperthite + aegirine + nepheline

(c) Nepheline + plagioclase + quartz

(d) Hypersthene + diopside + augite

115. What will be the AKF values of $Fe_3Al2Si_3O_{12}$?

(a) F = 75; A = 0; K =25 (b) F = 75; A = 25; K=0

(c) A = 75; F = 25; K =0 (d) K = 75; F = 25; A=0

116. Which one of the following minerals is formed during ultrahigh pressure metamorphism?

(a) Quartz (b) Tridymite

(c) Coesite (d) Labradorite

ANSWER SHEET

1. (c)	2. (d)	3. (d)	4. (a)	5. (b)
6. (c)	7. (d)	8. (d)	9. (d)	10. (c)
11. (a)	12. (a)	13. (b)	14. (a)	15. (c)
16. (c)	17. (b)	18. (a)	19. (c)	20. (c)
21. (a)	22. (b)	23. (b)	24. (a)	25. (c)
26. (a)	27. (b)	28. (c)	29. (b)	30. (a)
31. (c)	32. (c)	33. (b)	34. (c)	35. (b)
36. (b)	37. (c)	38. (d)	39. (d)	40. (b)
41. (a)	42. (d)	43. (a)	44. (d)	45. (b)
46. (c)	47. (a)	48. (a)	49. (b)	50. (b)
51. (a)	52. (c)	53. (c)	54. (d)	55. (b)
56. (c)	57. (a)	58. (d)	59. (a)	60. (d)
61. (a)	62. (b)	63. (b)	64. (c)	65. (b)
66. (d)	67. (d)	68. (b)	69. (c)	70. (b)
71. (c)	72. (b)	73. (c)	74. (a)	75. (c)
76. (d)	77. (d)	78. (b)	79. (c)	80. (a)
81. (c)	82. (c)	83. (b)	84. (a)	85. (c)
86. (c)	87. (a)	88. (c)	89. (a)	90. (b)
91. (d)	92. (c)	93. (b)	94. (d)	95. (a)
96. (b)	97. (a)	98. (a)	99. (c)	100. (b)
101. (c)	102. (d)	103. (c)	104. (a)	105. (b)
106. (c)	107. (c)	108. (a)	109. (b)	110. (c)
111. (b)	112. (d)	113. (d)	114. (b)	115. (b)
116. (c)				

(B) SEDIMENTARY PETROLOGY

1. **By which of the following are clay minerals represented?**
   (a) Resistates (b) Oxidates
   (c) Silicates (d) Hydrolysates

2. **Suite of igneous metamorphic and sedimentary rocks exposed in orogenic belt where oceanic lithosphere is emplaced at continental margin is called:**
   (a) Spilite (b) Kerotophyre
   (c) Ophiolite (d) Banakite

3. **Graded bedding and wave ripples are useful indicators of:**
   (a) Unconformities
   (b) Chronological order
   (c) Tops and bottoms of sedimentary layers
   (d) Non-conformity

4. **Which of the following represents secondary structures in sedimentary rocks?**
   (a) Burrows (b) Cross bedding
   (c) Concretions (d) Ripple Marks

5. **Gypsum and anhydrite are formed under which sedimentary environment?**
   (a) Euxinic facies (b) Black shale facies
   (c) Mollase facies (d) Evaporitic facies

6. **Which of the following can provide best information about the source rocks of a sandstone?**
   (a) Paleocurrent Analysis (b) Grain size
   (c) Lithic fragments (d) Weathering Index

7. **To be classified as 'laminae' the thickness of the layer should be –**
   (a) < 1 cm (b) 1-5 cm
   (c) 5-10 cm (d) >10 cm

8. Which of the following variety of limestone is entirely composed of organic detritus?

(a) Coquina (b) Encrinite

(c) Spergenite (d) Oolite

9. Fine-grained Sandstone has which of the following properties in comparison with clay?

(a) Higher porosity and lower permeability

(b) Higher porosity and higher permeability

(c) Lower porosity and lower permeability

(d) Lower porosity and higher permeability

10. Fragment of granite in the body of a sandstone are:

(a) Older than the sandstone

(b) Younger than the sandstone

(c) Of same age as the sandstone

(d) Generally older at times younger

11. The directional sedimentary structure is/are:

(a) Mud rock (b) Rain print

(c) Load cast (d) Cross bending

12. Mud cracks are used in the study of:

(a) Top and bottom of bed (b) Direction of current

(c) Depositional direction (d) Paleocurrent direction

13. Diagenesis is:

(a) Pre-depositional (b) Syn-depositional

(c) Syn-and post-depositional (d) Post-depositional

14. Which of the following is non-clastic?

(a) Siltstone (b) Shale

(c) Halite (d) Conglomerate

15. Normal salt (NaCl) is formed in which environment

(a) Igneous (b) Sedimentary

(c) Metamorphic (d) Volcanic

16. The principal mineral in most sand beaches is

(a) Quartz (b) Orthoclase
(c) Magnetite (d) Muscovite

17. Which of the following is the most important erosion and transportation agent?

(a) Wind (b) Water
(c) Glaciers (d) None of the above

18. Of the following minerals, the one most resistant to both chemical and mechanical weathering is:

(a) Calcite (b) Muscovite
(c) Hornblende (d) Quartz

19. When shale weather's, it forms mostly:

(a) Clay (b) Quartz
(c) Soluble carbonates (d) Slate

20. Most cave deposits from which of the following minerals?

(a) Hematite (b) Calcite
(c) Quartz (d) None of the above

21. In geological studies, a breccia contains particle that have following shape?

(a) Rounded (b) Angular
(c) Square (d) Spheroidal

22. The saltation of sand grains refers to the sand grains:

(a) Being coated with salt (b) Moving in jumps and bounces
(c) Moving in a steady stream (d) Being cemented together

23. Give the general terms used to designate the sand, silt and mud deposited by a stream, along its bank or upon it's floodplain, during periods of high water.

(a) Alluvium (b) Flood-loan
(c) Dump moraine (d) Dune deposit

24. The example of a rock whose minerals have been crushed into thin sheets or bands is:

(a) Shale (b) Schist
(c) Conglomerate (d) Granite

25. The hydrolysis or orthoclase results in the formation of:

(a) Shale (b) Kaolin
(c) Lime (d) Hydrochloric acid

26. All of the following are examples of chemically precipitated sedimentary rock except

(a) Gypsum (b) Limestone
(c) Sandstone (d) Dolomite

27. A rock composed for cemented sand and gravel is:

(a) Shale (b) Conglomerate
(c) Granite (d) Marble

28. When clay hardness into rock, it forms:

(a) Coal (b) Limestone
(c) Shale (d) Quartzite

29. Angular fragments have Petti John's Roundness grade between -

(a) 0 to 0.15 (b) 0.16 to 0.30
(c) 0.31 to 0.50 (d) 0.51 to 0.70

30. Which of the following forms due to chemical deposition

(a) Arkose (b) Breccia
(c) Limestone (d) Greywacke

31. Graded Bedding structure forms in -

(a) Dolomite (b) Stromatolite
(c) Sand Stone (d) Coral reef

32. Stylolite structure develops in-

(a) Limestone (b) Breccia
(c) Conglomerate (d) Boulder Bed

33. Which Name is used for 'Boulder Clay' -

(a) Greywacke (b) Till
(c) Dune (d) Arkose

34. The Contact Metamorphism of impure limestone causes crystallisation of which mineral?

(a) Wollastonite (b) Wolframite
(c) Barite (d) Fluorite

35. Which one of the following combinations facilitates soil formation?

(a) Weathering rate high and erosion rate low

(b) Weathering rate high and erosion rate high

(c) Weathering rate low and erosion rate low

(d) Weathering rate low and erosion rate high

36. Which of the following characterize the grain supported limestones?

(a) Wackestone (b) Pack stone

(c) Bound stone (d) Dolostone

37. Which of the following represents the initiation of ideal turbidite?

(a) Horizontally bedded siltstone with bioturbation

(b) Ripple drift cross laminated sandstones

(c) Graded Sandstone with sole marks

(d) Ripple bedded Sandstone inter-bedded with siltstone

38. Which of the following characterize the period of no deposition, no erosion and no perceptible rock record?

(a) Diastem (b) Unconformity

(c) Discontinuity (d) Anatoly

39. Which of the following pairs is not correct?

(a) Bird's eye structure : epigenetic structure

(b) Convolute bedding : Penecontemporaneous

(c) Stylolite : Dissolution

(d) Graded bedding : Syngenetic

40. Which of the following is the main distinguishing character between Arenites and Wackes?

(a) Arenites are texturally mature whereas Wackes are immature.

(b) Arenites are better sorted than Wackes

(c) Arenites contain rock fragments less than Wackes

(d) Arenites contain up to 15% matrix whereas Wackes contain above 15% matrix

41. Which sedimentary structures can be used to decipher top and bottom of beds?

(a) Graded bedding (b) Current ripples

(c) Deformational structures (d) All of the above

42. The composition of an arkose would be

(a) Quartz+ Lithics+(Matrix > 15%)

(b) Quartz+ Feldspars+(Matrix > 15%)

(c) Quartz+ Feldspars+(Matrix<15%)

(d) Quartz+ Lithics+(Matrix < 15)

43. An anthigenic growth forms during

(a) Sedimentation
(b) Diagenesis
(c) Palingenesis
(d) Anatexis

44. A turbidite deposit may show:

(a) Ripple marks and good sorting

(b) Current lamination and ripple marks

(c) Rain prints and flute marks

(d) Graded bedding and poor sorting

45. As per Wentworth's scale, clay size is defined as

(a) Less than 1/256 mm
(b) 1 mm to 1/256 mm
(c) More than 256 mm
(d) 0.5 mm to 1.5 mm

46. Micrite is

(a) A fine-grained carbonate
(b) A type of stalactite
(c) Similar to myrmekite
(d) A hexagonal iron carbonates

47. Rudite is a rock comprising

(a) Clay
(b) Silt
(c) Sand
(d) Pebbles

48. A rock composed mainly of quartz with matrix less than 15% is

(a) Quartz arenite
(b) Greywacke
(c) Litharenite
(d) Mudstone

49. Sabkha is a

(a) Supuratidal deposit
(b) Desert deposit
(c) Deep sea deposit
(d) River terrace deposit

50. A sedimentary rock may form by weathering and deposition of which rock type?

(a) Igneous/Sedimentary/Metamorphic

(b) Only sedimentary/Metamorphic

(c) Only Igneous/Metamorphic
(d) Only Igneous/Sedimentary

51. The Phi-scale is given by

(a) $\phi = - \log_2 d$
(b) $\phi = - \log_6 d$
(c) $\phi = - \log_{10} d$
(d) $\phi = - \log_2 (-d)$

52. What is the sand particle size?

(a) 2 to 0.092 mm
(b) 4 to 2 mm
(c) 2 to 0.0625 mm
(d) 2 to 0.0825 mm

53. Grain size analysis of classic rock cannot interpret

(a) Environment
(b) Source rock
(c) Energy flow
(d) Nature of transportation

54. The river sands are distinguished by

(a) Positive skewness
(b) Negative skewness
(c) High porosity
(d) Sand dunes

55. The ratio of number of grain-to-grain contacts to the total number of grains counted in the traverse is

(a) Packing density
(b) Packing proximity
(c) True porosity
(d) True density

56. The dimictic conglomerate are characterized by

(a) High porosity
(b) High matrix
(c) Low matrix
(d) Zero matrix

57. Endogenetic rocks are deposited due to

(a) Precipitation
(b) River deposition
(c) Volcanic ash
(d) Mechanical weathering

58. Graywacke is a type of rock

(a) Volcanic rock
(b) Non-elastic sedimentary
(c) Rudaceous sedimentary rock
(d) Arenaceous sedimentary rock

59. What is mineralogical maturity of sandstone?

(a) Ratio of quartz to the olivine
(b) Ratio of feldspar to the quartz

(c) Ratio of quartz to feldspar
(d) Ratio of flakey minerals to feldspar

60. At 20° temperature, what will critical Reynolds?

(a) 500 (b) 1200
(c) 700 (d) 1000

61. Calcium carbonate could not found in

(a) Above CCD (b) Below CCD
(c) Below ACD (d) Above ACD

62. Which of the following original components bound together at time deposition?

(a) Mudstone (b) Wackestone
(c) Packatone (d) Boundstone

63. Which of the following statement is false?

(a) Graded bedding show's the grain size variation within a bed
(b) Convolute bedding a deformed bedding structure
(c) Secondary sedimentary structures form before the digenesis of a sedimentary rock
(d) Hummocky cross stratification is primary sedimentary structure

64. What is the main difference between a conglomerate and brescia?

(a) Brescia clasts are angular, conglomerate clasts are rounded
(b) A brescias is well stratified, a conglomerate is poorly stratified
(c) Brescia clasts are the size of baseballs, conglomerate claste are larger
(d) Brescia has a compacted, clay rich matrix conglomerate has no matrix

65. Detrital sedimentary rocks are classified based on the

(a) Colours of the cementing minerals
(b) Grain sizes of the detrital particles
(c) Grain sizes of the detrital particles
(d) Compositions of soluble minerals

66. Which of the following sedimentary rocks indicate long distance transportation of the sediments?

(a) Quartz arenite
(b) Brescia
(c) Arkose (sandstone with lots of feldspar particles)
(d) Felsparthic wacke

67. Which one is an argillaceous rock?

(a) Sandstone (b) Limestone
(c) Conglomerate (d) Shale

68. Large-scale cross-beds have the minimum thickness of

(a) 10 cm (b) 50 cm
(c) 66 cm (d) 5 cm

69. Debris flow produce

(a) Clast supported texture (b) Grain supported texture
(c) Matrix supported texture (d) Cement supported texture

70. Presence of mineral glauconitic suggests:

(a) Desert environment (b) Glacial environment
(c) Fluvial environment (d) Marine environment

71. Facies association of a prograding delta is

(a) Coarsening upward
(b) Fining upward
(c) Disorganized
(d) Both fining upward and disorganized

72. Greywacke sandstone indicated:

(a) Active provenance tectonics and prolonged transport
(b) Stable provenance and prolonged transport
(c) Active provenance tectonics and less transport
(d) Stable provenance and less transport

73. Presence of mineral Glauconite suggests

(a) Desert environment (b) Glacial environment
(c) Fluvial environment (d) Marine environment

74. The red shale or Arkosic facies are

(a) Marine one
(b) Both Marine and Non-marine
(c) Non-marine one
(d) Metamorphic one

75. Greywacke sandstone indicates

(a) Active provenance tectonics and prolonged transport
(b) Stable provenance and prolonged transport
(c) Stable provenance and less transport
(d) Active provenance tectonics and less transport

76. Chromite deposits are mostly of

(a) Igneous origin
(b) Metamorphic origin
(c) Sedimentary origin
(d) Both Metamorphic and Sedimentary origin

77. Which one of the following is a diagenetic process?

(a) Metamorphism
(b) Dissolution
(c) Deposition of sediments
(d) Palingenesis

78. Which one of the following is a unidirectional paleo current indicator?

(a) Grovecasts
(b) Symmetrical ripples
(c) Channel and scour margins
(d) Cross- bedding

79. Which of the following are the lithologies of deltaic deposits?

(a) Siltstone, greywacke and claystone
(b) Conglomerate, sandstone and mudstone
(c) Shale, clay and limestone
(d) Limestone, dolomite and shale

80. Which one of the following is NOT a clastic rock?

(a) Sandstone
(b) Shale
(c) Limestone
(d) Conglomerate

81. A geothermometer is based on which one of the following type of reactions?

(a) Net transfer reaction
(b) Ion exchange reaction
(c) Oxidation reaction
(d) Dehydration reaction

82. Which one of the following rocks is the protolith for marble?

(a) Arenaceous rocks (b) Argellaceous rocks
(c) Mafic rocks (d) Calcareous rocks

83. The organic compounds in sediments, rocks and crude oil whose carbon structures can be traced back to a living organism are called as:

(a) Bio markers (b) Fossil markers
(c) Trace markers (d) Index markers

84. A feldspar-rich sandstone is called a(n) ________.

(a) Arkose (b) Litharenite
(c) Quartz arenite (d) Shale

ANSWER SHEET

1. (d)	2. (c)	3. (c)	4. (a)	5. (d)
6. (b)	7. (a)	8. (d)	9. (b)	10. (a)
11. (d)	12. (a)	13. (c)	14. (c)	15. (b)
16. (a)	17. (b)	18. (d)	19. (a)	20. (b)
21. (b)	22. (b)	23. (a)	24. (c)	25. (b)
26. (c)	27. (b)	28. (c)	29. (a)	30. (c)
31. (c)	32. (a)	33. (b)	34. (a)	35. (a)
36. (a)	37. (b)	38. (b)	39. (b)	40. (d)
41. (a)	42. (c)	43. (b)	44. (d)	45. (a)
46. (a)	47. (d)	48. (a)	49. (a)	50. (a)
51. (a)	52. (c)	53. (b)	54. (a)	55. (b)
56. (b)	57. (a)	58. (d)	59. (c)	60. (a)
61. (b)	62. (d)	63. (c)	64. (a)	65. (d)
66. (a)	67. (d)	68. (d)	69. (c)	70. (d)
71. (a)	72. (c)	73. (d)	74. (c)	75. (d)
76. (a)	77. (b)	78. (d)	79. (b)	80. (c)
81. (b)	82. (d)	83. (a)	84. (a)	

(C) METAMORPHIC PETROLOGY

1. **During metamorphic recrystallization, the free energy between the crystal faces is minimized by the development of crystal faces at an angle of about:**
   (a) 45° (b) 60°
   (c) 90° (d) 100°

2. **Which of the following is the diagnostic mineral for very low grade metamorphic?**
   (a) Epidote (b) Laumonite
   (c) Clinozoisite (d) Lawsonite

3. **Schorl rock is formed by:**
   (a) Sulphur metasomatism (b) Boron metasomatism
   (c) CO_2 metasomatism (d) Chlorine metasomatism

4. **Which of the following mineral sequences depicts the increasing metamorphic grade of pelitic rocks?**
   (a) Biotite – kyanite – garnet – staurolite – chlorite
   (b) Chlorite – garnet – biotite – staurolite – kyanite
   (c) Chlorite– biotite – garnet – staurolite – kyanite
   (d) Garnet – biotite – chlorite – kyanite – staurolite

5. **Which of the following amphibole is characteristic of high pressure-low temperature metamorphic conditios?**
   (a) Hornblende (b) Tremolite
   (c) Actinolite (d) Glaucophane

6. **Which of the following rocks is a product of contact metamorphism?**
   (a) Lamprophyre (b) Charnockite
   (c) Granulite (d) Skarn

7. **Which of the following represent lineation?**
   (a) Intersection of two sets of cleavages
   (b) Parallel arrangement of elongated minerals
   (c) Mullions
   (d) All the above

8. Straty cleavage, schistosity and gneissose are characteristically found in:

(a) All type of igneous rocks
(b) All type of sedimentary rocks
(c) All type of metamorphic rocks
(d) Only regionally metamorphosed rocks

9. The metamorphic rock which forms from pure limestone is:

(a) Slate (b) Gneiss
(c) Schist (d) Marble

10. Which coal is considered a metamorphic rock?

(a) Anthracite coal (b) Bituminous coal
(c) Lignite coal (d) All of these

11. Which of the following is a metamorphic to rock?

(a) Sandstone (b) Granite
(c) Quartz (d) Marble

12. Of the following the rock that is most resistant to both chemical and mechanical weathering is:

(a) Shale (b) Limestone
(c) Marble (d) Quartzite

13. Match List-I with List-II and select the correct answer using the codes given below the Lists:

	List-I (Rock)		List-II (Texture)
A.	Amphibolite	1.	Cataclastic
B.	Schist	2.	Granoblastic
C.	Charnockite	3.	Lepidoblastic
D.	Mylonite	4.	Nematiblastic

Code:

	A	B	C	D
(a)	1	2	3	4
(b)	4	2	3	1
(c)	1	3	2	4
(d)	4	3	2	1

14. Which one of the following implies 'Helicitic structure'?

(a) Any arrangement of inclusions representing a relict structure preserved from an earlier metamorphism

(b) Arrangement of inclusions in concentric layers in a porphyroblast

(c) Development of concentric zone or zones of one or more minerals surrounding another mineral

(d) Relicts of an earlier mineral resulting from its replacement during metamorphism

15. Which one of the following associations is rare in metamorphosed pelitic rocks?

(a) Cordierite - Sillimanite - Plagioclase

(b) Garnet - Cordierite - Plagioclase

(c) Cordierite - Kyanite - Garnet

(d) Cordierite - Sillimanite - Garnet

16. Consider the following statements:

1. Low pressure metamorphic belts show a transaction from Andalusite to sillimanite.
2. In medium pressure belts kyanite converts to sillimanite.
3. Garnet takes the place of chlorite and relatively low-pressure phase Cordierite shows up at this level.
4. High pressure belts contain dense phases of jadeite or glaucophane.

Which of these statements characterize the assembles developed on account of metamorphism of pelitic rocks?

(a) 1, 2 and 4 (b) 1 and 3 only

(c) 2 and 4 only (d) 2, 3 and 4

17. Which one of the following is not a prograde reaction?

(a) Dolomite + Quartz —— Talc + Calcite

(b) Wollastonite + CO_2 —— Calcite + Quartz

(c) 3 Dolomite + Diaspore — 4 Calcite + 2 Forsterite + 2 CO_2

(d) Tremolite + 3 Calcite + 2 Quartz — 5 Diopside + $3CO_2$ + H_2O

18. Omphacite, characteristically found in eclogite is a

(a) Sodic hornblende (b) Sodic pyroxene

(c) Sodic garnet (d) Sodic plagioclase

19. Texture formed by parallel arrangement of acicular minerals in a metamorphic rock is called:

(a) Crystalloblastic (b) Nematoblastic
(c) Lepidoplastic (d) Porphyroblastic

20. Saccharoidal texture is seen in which rock?

(a) Syenite (b) Hornfels
(c) Marble (d) Biotite schist

21. An AFM diagram depicts the metamorphic mineral assemblage of

(a) Politic rocks (b) Charnockites
(c) Hornfels (d) Greenstones

22. The beginning of metamorphism is indicated by the first appearance of

(a) Staurolite (b) Lawsonite
(c) Quartz (d) Kyanite

23. Lauontite is stable at water pressure of

(a) > 7 kb (b) 7 to 5 kb
(c) 5 to 3 kb (d) < 3 kb

24. Vesuvianite is formed in

(a) Metamorphosed arenites (b) Metamorphosed pelites
(c) Metamorphosed marls (d) Metamorphosed carbonates

25. Texture of charnockite is

(a) Granoblastic (b) Porphyritic
(c) Hypidiomorphic (d) Lepidoblastic

26. With increasing metamorphism, sequence of index minerals will be

(a) Biotite-Chlorite-Kyanite
(b) Kyanite-Sillimanite-Almandine
(c) Almandine-Staurolite-Kyanite
(d) Biotite-Sillimanite-Kyanite

27. Muscovite + Quartz = K–feldspar + Al_2SiO_5 + H_2O (water pressure < 3.5 kb) defines:

(a) Beginning of very low-grade metamorphism
(b) Beginning of low-grade metamorphism
(c) Beginning of medium grade metamorphism
(d) Beginning of high-grade metamorphism

28. The metal content in an ore is called as:

(a) Grade (b) Tenor
(c) Rank (d) Gangue

29. Which one among the following is not metamorphic rock?

(a) Hornfels (b) Orthoquartzite
(c) Metgranite (d) Biotitic gneiss

30. Which one of the following belongs to high grade metamorphism?

(a) Eplzone (b) Acme zone
(c) Mesozone (d) Katazone

31. ……… structure is found in the rocks composed of equidimensional minerals:

(a) Gneissose (b) Maculose
(c) Granulose (d) Schistose

32. Which rock is a para metamorphic?

(a) Metacharnokite (b) Metaqurtsite
(c) Biotitic gneiss (d) Khondalite

33. Which one from the following is non-folisted metamorphic rock?

(a) Gneiss (b) Phyllite
(c) Slate (d) Quartzite

34. …………. is the high temperature and high-pressure metamorphic facies.

(a) Blueschist (b) Greenschist
(c) Ecologies (d) Amphibolite

35. The term 'blast or blastic' used as prefix in the case of

(a) Palimpsest texture (b) Aphanitic texture
(c) Phanitic texture (d) Glassy texture

36. In the Al_2SiO_5 polymorphs, which shows highest grade of metamorphism?

(a) Andalusite (b) Kyanite

(c) Sillimanite (d) All of the above

37. In the following indicates the ascending order of temperature of formation

(a) Eclogite, Greenschist, Zoolite, Blueschist

(b) Greenschist, Blueschist, Eclogite, Granulite

(c) Zeolite, Amphibolite, Granulite, Eclogite

(d) Blueschist, Granulite, Amphibolite, Eclogite

38. Omphacite and pyrope garnet are the representative minerals for

(a) Zoolite facies (b) Greenschist facies

(c) Glaucophane-schist facies (d) Eclogite facies

39. Which is not related to metamorphism?

(a) Wind (b) Pressure

(c) Temperature (d) Chemically active fluid

40. A uniform pressure is also called

(a) Directed pressure (b) Hydrostatic pressure

(c) Compression (d) Tension

41. .... metamorphism occurs around larger intrusive at comparatively low temperature.

(a) Pyro metamorphism (b) Contact metamorphism

(c) Metasomatism (d) Auto metamorphism

42. Match the following Columns and give the correct answer:

	Column-A		Column-B
P.	Blueschist faciea	1.	Chlorite
Q.	Greonschist	2.	Plagioclase
R.	Granulite	3.	Mg-Garnet
S.	Eclogite	4.	Glaucophane

(a) P-4, Q-1, R-2, S-3 (b) P-1, Q-4, R-3, S-2

(c) P-2, Q-3, R-1, S-4 (d) P-3, Q-2, R-4, S-1

43. Correctly match column A and B

	Column-A		Column-B
P.	Shale	1.	Quartzite
Q.	Sandstone	2.	Amphibolites
R.	Granite	3.	Slate
S.	Basalt	4.	Gneiss

(a) P-1, Q-2, R-4, S-3
(b) P-2, Q-1, R-4, S-3
(c) P-3, Q-1, R-4, S-2
(d) P-4, Q-3, R-2, S-1

44. Find odd one out

(a) Granite
(b) Basalt
(c) Slate
(d) Diorite

45. Which one is the completely infoliated rock?

(a) State
(b) Schist
(c) Phyllite
(d) Hornfels

46. Pyro metamorphism generally includes:

(a) High pressure changes
(b) High temperature changes
(c) Low pressure and temperature
(d) High pressure and temperature changes

47. Migmatites are the result of

(a) Retrograde metamorphism
(b) Ultra metamorphism
(c) Palingenesis
(d) Metasomatism

48. Khondalites are characteristic rocks of

(a) Amphibolites facies
(b) Granulite facies
(c) Eclogite facies
(d) Green-schist facies

49. Find odd one out

(a) Marble
(b) Slate
(c) Granite
(d) Phyllite

50. Tormalinisation is a combined effect of

(a) Water and fluorine
(b) Water, borone and fluorine

(c) Water, carbon dioxide and chlorine
(d) Water, carbon, dioxide and hydrogen

51. Thermal metamorphism of shales produces

(a) Hornfels (b) Novaculite
(c) Phyllite (d) Schist

52. The metamorphic rock essentially composed of hornblende and plagioclase is

(a) Amphibolite (b) Hornblendite
(c) Blue schist (d) Hornfels

53. The most comman accessory mineral in eclogites is

(a) Ilmenite (b) Zoisite
(c) Rutile (d) Sphene

54. Which one is a metamorphic texture?

(a) Ophitic (b) Clastic
(c) Granoblastic (d) Subophitic

55. The metamorphic rock with maculose structure is

(a) Granulose (b) Hornfels
(c) Cataclastic (d) Schistose

56. Find odd one out

(a) Idioplasmic (b) Non-clastic
(c) Poikilitic (d) Blastoporphyritic

57. Regional metamorphism:

(a) Produces metamorphic rocks over large areas
(b) Produces Foliated metamorphic rocks
(c) Is occurring as you read these questions
(d) All of these

58. Slate is a metamorphic rock type associated with:

(a) Low temperature, high pressure conditions
(b) High temperature, low pressure conditions
(c) Low temperature, low pressure conditions
(d) High temperature, high pressure conditions

59. Gneiss is a metamorphic rocks associated with:

(a) High temperature, high pressure conditions
(b) High temperature, low pressure conditions
(c) Low temperature, low pressure conditions
(d) Low temperature, high pressure conditions

60. Metamorphism of limestone may contribute to global warming by the release of:

(a) Oxygen
(b) Sulfuric acid
(c) Carbon dioxide
(d) Nitrous oxide

61. Migmatites are the result of

(a) Ultra metamorphism (b) Retrograde metamorphism
(c) Palingenesis (d) Metasomatism

62. Khondalites are characteristic rocks of

(a) Amphibolite facies (b) Eclogite facies
(c) Granulite facies (d) Green-schist facies

63. A line joining the points where the rocks have the same grade of metamorphism is called

(a) Isogyre (b) Isograde
(c) Isopach (d) Isochore

64. Find odd one out

(a) Marble (b) Slate
(c) Granite (d) Phyllite

65. The typical product of contact metamorphism with maculose structure

(a) Granulose (b) Cataclastic
(c) Hornfels (d) Schistose

66. The most common accessory mineral in eclogites is

(a) Ilmenite (b) Zoisite
(c) Rutile (d) Sphene

ANSWER SHEET

1. (c)	2. (d)	3. (b)	4. (b)	5. (d)
6. (d)	7. (d)	8. (d)	9. (d)	10. (d)
11. (d)	12. (d)	13. (c)	14. (b)	15. (a)
16. (b)	17. (c)	18. (b)	19. (b)	20. (c)
21. (a)	22. (b)	23. (d)	24. (c)	25. (a)
26. (c)	27. (d)	28. (b)	29. (b)	30. (d)
31. (c)	32. (b)	33. (d)	34. (c)	35. (a)
36. (c)	37. (c)	38. (d)	39. (a)	40. (b)
41. (b)	42. (a)	43. (c)	44. (c)	45. (d)
46. (b)	47. (b)	48. (b)	49. (c)	50. (b)
51. (a)	52. (a)	53. (c)	54. (c)	55. (b)
56. (c)	57. (d)	58. (d)	59. (c)	60. (a)
61. (a)	62. (c)	63. (b)	64. (c)	65. (c)
66. (c)				

CHAPTER - 5

STRUCTURAL GEOLOGY

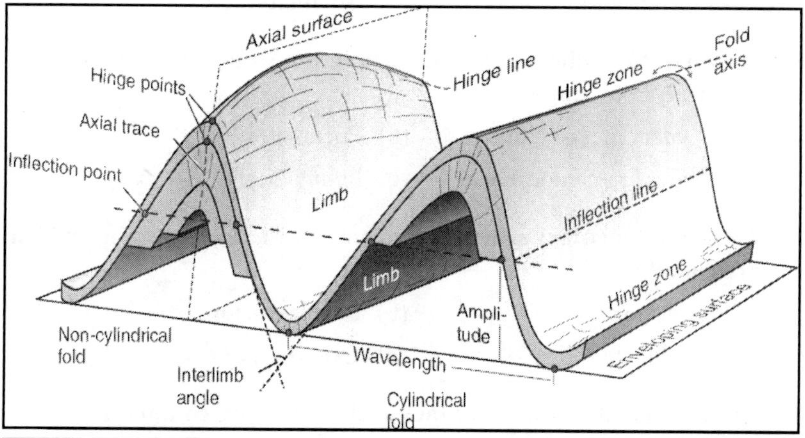

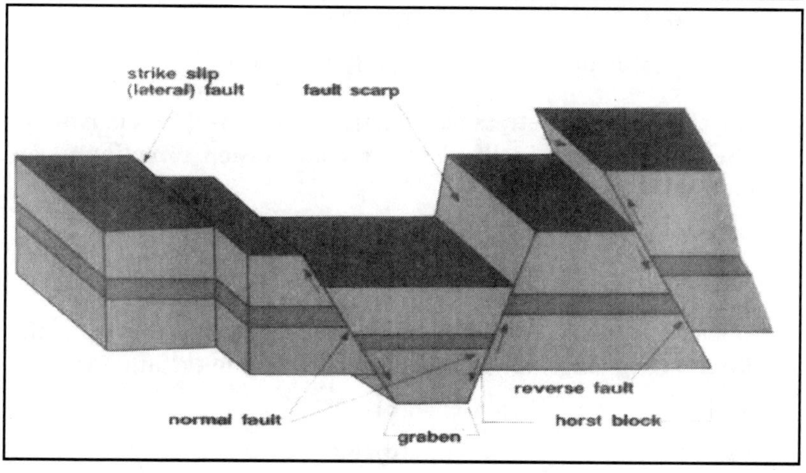

Structural Geology

1. **If the cleavage is dipping more gently than the bedding plane, the beds are said to be:**

   (a) Horizontally turned (b) Right side up

   (c) Right side down (d) Upside down

2. **When dip of the beds of a fold are almost parallel except near the hinge line, the folds are called:**

   (a) Open folds (b) Inclined folds

   (c) Parallel folds (d) Isoclinal folds

3. **Which of the following processes is responsible for ptygmatic of a quartz vein in amphibole?**

   (a) Contact metasomatism (b) Pneumatolysis

   (c) Optalic metamorphism (d) Palingenesis

4. **Two complimentary sheer planes constitute two sets of joints which are called:**

   (a) Shear joints (b) Sheet structure

   (c) Conjugate joints (d) Extension joints

5. **In extensional tectonic regimes of the Earth we commonly**

   (a) Normal fault (b) Thrust

   (c) Vertical fault (d) Reverse fault

6. **The effective shear stress on a frictional plane will be a maximum when the plane is inclined to the compression direction at an angle of**

   (a) 90° (b) >45°

   (c) <45° (d) 45°

7. **In a Metric toposheet published by the SOI on 1:50,000 scale, the difference in longitude/latitude from one end to the other is:**

   (a) 5° (b) 10°

   (c) 15° (d) 30°

8. **A person standing at 90°00' N latitude uses a Brunton compass as a guide and travels 320 Km due South then 450 Km East and finally 320 Km North. His final position will be near:**

   (a) 90°00'N (b) 54°00'N:52°35'E

   (c) 74°15'N:52°35'E (d) 80°12'N:52°35'E

9. **An instrument used for studying a rock sample side in different orientation is called:**

   (a) Point counter (b) Dollar counter

   (c) Integrating stage (d) Universal stage

10. **Which of the following rocks sample can be used as an oriented sample?**

    (a) Sample is a small fold. Data-axial plane dip 25°

    (b) Foliated rock from 26°30'N:72°55'E at 1588 m on top of hill

    (c) Limestone with data dip 035°/315° from 28°40'N:78°15'E

    (d) Sandstone with markings-N28°E; and two sides with Level marks

11. **During section measurement it is found a bed dips 30° due north, the slope of the ground is 20° due North then**

    (a) Outcrop width will be less than true thickness of bed

    (b) Outcrop width will be equal to the true thickness of bed

    (c) Outcrop width will be more than true thickness of bed

    (d) Outcrop width will always be less than the true thickness

12. **In an inclined bed the true dip amount is 25°, the amount of dip in strike direction will be:**

    (a) 0° (b) 10°

    (c) 20° (d) 25°

13. **In the core of a syncline the beds**

    (a) Can be older or younger (b) Are older

    (c) Are younger (d) Are same as in an anticline

14. **In a recumbent fold the axial plane dip nearly at**

    (a) 90° (b) 45°

    (c) 25° (d) 0°

15. The area of a degree sheet is divided into how many toposheets?

(a) 4
(b) 8
(c) 16
(d) 32

16. On a topographic map, 4 cm on the map is equivalent to 2.0 km on ground. What is the RF?

(a) 1:500
(b) 1:50000
(c) 1:5000
(d) 1:50

17. Folds with a pair of converging axial surface and having a square hinge zone are known as

(a) Parallel
(b) Similar folds
(c) Box folds
(d) Overturned folds

18. Low angle normal fault are called

(a) Thrusts
(b) Lags
(c) Horst and Graben
(d) Shear zones

19. Imaginary line joining points of equal elevation on the surface of the Earth's is called

(a) Isograd line
(b) Contour line
(c) Isobaric line
(d) Strike line

20. A fracture in a rock with no appreciable movement is called-

(a) Slickenside
(b) Gash vein
(c) Joint
(d) Shear

21. Transform fault is

(a) Strike slip fault
(b) Dip slip fault
(c) Up slip fault
(d) Down slip fault

22. According to principle of cross-cutting relationship, intrusive body is

(a) Older than country rock
(b) Younger than country rock
(c) Of same age as the country rock
(d) None of these

23. On a topographic Mao, lines that connect points of the same elevation are termed:

(a) Benchmarks (b) Bathymetric lines

(c) Contour lines (d) Isobars

24. Of the following, the colour blue on a topographic map indicates what surface feature ?

(a) Vegetation (b) Water

(c) Mountain (d) Rock formation

25. Which of the following term's is used to describe the angular difference between TRUE north and magnetic north on a topographic map?

(a) Magnetic field (b) Magnetic intensity

(c) Magnetic inclination (d) Magnetic declination

26. In geological studies, a fault in the Earth's surface in which the hanging wall has moved UP relative to the footwall is termed:

(a) Overthrust fault (b) Reverse fault

(c) Test fault (d) None of these

27. In geological studies, are fault in the Earth's surface along which primarily horizontal movement has taken place is termed:

(a) Overthrust fault (b) Strike-slip-fault

(c) Reverse fault (d) None of these

28. In studies of the surface of the earth, a reverse fault having a dip between 10 and 45 degrees is known as a:

(a) Thrust fault (b) Normal fault

(c) Strike-slip fault (d) None of these

29. What is the name for an erosion surface that separates two sets of sedimentary layers with non-parallel bedding planes?

(a) Cross bedding (b) Graded bedding

(c) Fault (d) Angular unconformity

30. Which of the following could be used to tell if a rock layer is right side up or up side down?

(a) Graded bedding (b) Faults

(c) Lineation (d) None of above

31. Principal of superposition states that in an underformed sedimentary sequence a bed is _____than the bed above and _____than the bed below:

(a) Younger, younger (b) Younger, older

(c) Older, younger (d) Older, younger

32. A fold with parallel arrangement of limbs is:

(a) Isoclinal fold (b) Inverted fold

(c) Symmetrical fold (d) Parallel fold

33. A fold which is convex upwards and have younger rocks in it's core is:

(a) Antiform (b) Synform

(c) Anticlinorium (d) Anticline

34. Fold having sharp hinges and smooth limbs are called

(a) Concentric (b) Cuspate

(c) Chevron (d) Conical

35. The intersection of contour lines with bedding planes indicate the presence of:

(a) Horizontal beds (b) Inclined beds

(c) Unconformity (d) Graded beds

36. A clinometer compass is used to measure:

(a) Strike direction

(b) Strike and dip direction

(c) Strike, dip direction and dip amount

(d) None of these

37. Dip of bed measured perpendicular to strike direction is called:

(a) Normal dip (b) False dip

(c) True dip (d) Apparent dip

38. True dip is always more than:

(a) Normal dip (b) False dip

(c) Apparent dip (d) None of above

39. Which of the following is not an example of brittle deformation?

(a) Fold (b) Joint

(c) Fault (d) Fracture

40. Which of the following processes can lead to the formation of faults?

(a) Compression (b) Extension

(c) Shearing (d) All the above

41. Head of fault is the

(a) Angle between fault plane and bedding plane

(b) Angle between fault plane and horizontal plane

(c) Angle between hanging wall and foot wall

(d) Angle between fault plane and vertical plane

42. Which of the following statement is true for Horst and Graben structure?

(a) They are bound by normal faults on either side

(b) They are non-tectonic in origin

(c) They are formed by differential weathering of basaltic rocks

(d) They are formed under compressive forces

43. Columnar joints are generally associated with

(a) Contact metamorphic rocks

(b) Sedimentary rocks of aeolian origin

(c) Volcanic rock

(d) Plutonic rock

44. Which of the following statement is true for slaty cleavage?

(a) These are closely spaced foliations

(b) These are associated with deformation of rocks

(c) They are generally parallel to the axial planes of folds

(d) All of above

45. Fault with no vertical movement is called

(a) Thrust fault (b) Transform fault

(c) Reverse fault (d) Normal fault

46. Rocks formed in layers are said to be:

(a) Crystalline (b) Stratified
(c) Chemical (d) Metamorphosed

47. Which of the following statements is correct for reverse faults?

(a) The Foot Wall moves up relative to the Hanging Wall
(b) There is no component of vertical movement
(c) The Hanging Wall moves up relative to Foot Wall
(d) The Hanging Wall moves down relative to Foot Wall

48. If both the limbs of a fold dip in the same direction and the axial plane is inclined, the fold can be called as –

(a) Monoclinal fold (b) Chevron fold
(c) Reclined fold (d) Isoclinal fold

49. In which type of folding the thickness of beds remains constant?

(a) Similar (b) Concentric
(c) Parallel (d) Cylindrical

50. In a 1:50,000 scale Topographic map a distance of 0.25 cm represents-

(a) 1250 m (b) 250 m
(c) 125 m (d) 500 m

51. A limestone bed dips at 36° in N 60° E direction. What will be the dip of this limestone bed when measured in a vertical plane trending N 30°W?

(a) 90° (b) 0°
(c) 36° (d) 63°

52. When fold axis and axial plane of the fold; both dip steeply, the fold in named as –

(a) Box fold (b) Reclined fold
(c) Kink band (d) Monocline

53. In parallel fold geometry –

(a) The fold hinge is thick and limbs are thin
(b) The fold hinge is thin and limbs are thick
(c) The fold hinge and limbs both have same thickness

(d) The interlimb angle is more than 110° hinge is thicker than limbs

54. Columnar joint structure can generally be seen in one of the following rocks

(a) Gabbro (b) Basalt

(c) Granite (d) Syenite

55. Pitch is defined as-

(a) Acute angle between strike line and linear structure

(b) Obtuse angle between strike line and linear structure

(c) Plunge direction of a linear structure

(d) Plunge angle measure of a linear structure

56. Pitch is measured with the help of -

(a) Stereoscope (b) Protractor

(c) Compass (d) Tape

57. Which one of the following is a primary structure?

(a) Joint plane (b) Fault plane

(c) Bedding plane (d) Cleavage plane

58. The line of intersection of two planes pitches at 90° on both the planes, which of the following interpretation is correct?

(a) Both planes are vertical

(b) The strikes of the two inclined planes are manually perpendicular

(c) One plane is horizontal

(d) One plane is vertical

59. Which one of the following expressions donates maximum inclination of an inclined bed?

(a) 1:25 (b) 1:40

(c) 1:10 (d) 1:30

60. A set of horizontal beds overlie another set of horizontal beds separated by an irregular unconformity surface. What is the type of unconformity?

(a) Non-conformity (b) Para conformity

(c) Angular unconformity (d) Disconformity

61. Which one of the following folds is characterized by a constant orthogonal thickness in all the parts of the fold?

(a) Similar fold (b) Parallel fold

(c) Flow fold (d) Shear fold

62. Which one of the following term describes the joints that are perpendicular to the regional fold axis?

(a) Cross-joints (b) Columnar joints

(c) Hybrid joints (d) Shear joints

63. Which one of the following represents a curved fault plane that dips gently in depth and becomes steeper nearer to the surface?

(a) Rotational fault (b) Reverse fault

(c) Listric fault (d) Hinge fault

64. Which one of the following is correct while taking bearing of an object with a clinometer compass?

(a) Pointing N end of the magnetic needle towards the object

(b) Pointing N marked by the crown on the dial of the compass towards the object

(c) Pointing S end of the needle towards the object

(d) Pointing E or W marked on the dial towards the object

65. Which one of the following deposits is characterized by herring bone cross bedding?

(a) Fluvial (b) Tidal

(c) Delta (d) Shelf

66. The stereographic projection of poles to hedding surfaces is found to lie on a common great circle. Which one of the following is the correct interpretation of the structure?

(a) Non-cylindrical folding (b) Cylindrical folding

(c) Conical folding (d) Isoclinal folding

67. Point of maximum curvature in fold is defined as

(a) Hinge point (b) Inflection point

(c) Plunge (d) Centre of curvature

68. Tran current fault is a type of

(a) Normal fault

(b) Reverse fault

(c) Stike slip fault

(d) Oblique fault

69. A plane is dipping 45° (True dip) towards NE, the apparent dip of plane can be

(a) >90°
(b) Only 0°
(c) 45° to 0°
(d) 90° to 45°

70. True dip of a plane is 30°, what will be head of the plane?

(a) 45°
(b) 60°
(c) 90°
(d) 0°

71. The direction of line along which an inclined bed or plane intersects a horizontal plane is known as:

(a) Fault trace
(b) Dip direction
(c) Strike
(d) Plunge

72. Point of maximum elevation in fold is termed as

(a) Hinge point
(b) Tip point
(c) Crest
(d) Tough

73. Structural Geology deals with matters in the:

(a) Solid state
(b) Liquid state
(c) Gaseous state
(d) All of the above

74. Unconformity is define as:

(a) Plane representing the time of non-deposition and erosion

(b) Plane of asymmetric deposition

(c) Plane of joint and fault on regional scale

(d) Plane of faulting

75. Which one is not a deformational structure?

(a) Fold
(b) Fault
(c) Joint
(d) Disconformity

76. Match the following

	Group I (Terms)		Group II (Classes)
P.	Cross section	1.	Dip isogones that are perpendicular to bedding throughout the fold
Q.	Profile section	2.	A vertical plane through a fold
R.	Similar fold	3.	Dip isotones that are parallel to each other
S.	Parallel fold	4.	The surface perpendicular to the hinge line

(a) P-1, Q-2, R-4, S-3
(b) P-4, Q-2, R-1, S-3
(c) P-4, Q-2, R-3, S-1
(d) P-1, Q-2, R-3, S-4

77. Intermediate principal stress is vertical in

(a) Normal fault
(b) Strike slip fault
(c) Reverse fault
(d) Thrust fault

78. Which of following is a correct statement?

(a) The angle between fold limbs as measured in the cross section is called the interlimb angle
(b) A fold with a horizontal axial surface by definition must have a horizontal hinge line, and is called a reclined fold
(c) The high point of the hinge line in a doubly-plunging fold is called the culmination
(d) In an anticline, the beds are younger towards the core

79............ is an area of older rocks surrounded by younger rocks.

(a) Inlier
(b) Outlier
(c) Dome
(d) Basin

80. Having net slip equal to dip slip and rake of net slip 900.

(a) Dip fault
(b) Dip slip fault
(c) Strike fault
(d) Bedding fault

81. .... is defined as a fracture in a rock between the sides of which there is no observable movement.

(a) Fold
(b) Fault
(c) Joint
(d) Shear Zone

82. A topographic contour is defined as

(a) Line joining the point of equal stratigraphic
(b) Line joining the point of equal elevation
(c) Line joining the point of equal thickness
(d) Line joining point of same age rocks

83. A rock cut slope having inclination 40° due east the joint set most likely to cause failure is the one dipping at

(a) 20° due 0
(b) 30° due E
(c) 60° due E
(d) 65° due E

84. Amount of dip is maximum

(a) In the direction of strike
(b) At 90° to the strike direction
(c) At 45° to the strike direction
(d) At 0° to the strike direction

85. In a recumbent fold the axial plane is nearly

(a) Horizontal
(b) Vertical
(c) Inclined at 45°
(d) Inclined at 30°

86. The average density of the earth is:

(a) 4.5 gm/cm^3
(b) 5.5 gm/cm^3
(c) 6.5 gm/cm^3
(d) 7.5 gm/cm^3

87. Volcanic activity is usually absent along.

(a) Divergent plate boundaries
(b) Convergent plate boundaries
(c) Transform faults
(d) All of the above

88. A strike line with a given value is called:

(a) Horizontal equivalent
(b) Contour interval
(c) Stratum contour
(d) Contour line

89. Columnar joints are seen in:

(a) Basalts
(b) Sandstones
(c) Gneissic rocks
(d) Pegmatites

90. Boudins are formed in:

(a) Extensional regimes
(b) Compressional regimes
(c) Zero stress regimes
(d) Fault regimes

91. Pitch is another term for:

(a) Hade (b) Rake
(c) Plunge (d) Axial angle

92. An imaginary line which joins points of equal elevation is called:

(a) Contour line (b) Isothermal line
(c) Isosesmal line (d) Isograde line

93. Folds with inter-limb angle > 70º are called:

(a) Open folds (b) Closed folds
(c) Tight folds (d) Isoclinals folds

94. Gouge is associated with:

(a) Joints (b) Folds
(c) Unconformities (d) Faults

95. Folds having parallel limbs are:

(a) Parallel folds (b) Similar folds
(c) Isoclinal folds (d) Infinite folds

96. Dip of the bed is always measured in

(a) Horizontal plane (b) Vertical plane
(c) Inclined plane (d) Axial plane

97. An unconformity with almost parallel beds overlying and underlying the surface of erosion is

(a) Non-conformity (b) Disconformity
(c) Para-unconformity (d) Angular unconformity

98. The folds with thickened crests or troughs and thinner limbs are

(a) Open folds (b) Closed folds
(c) Isoclinals folds (d) Over folds

99. Abrupt termination of strata marks the presence of

(a) Fold and joint (b) Joint
(c) Fold (d) Fault

100. The folds having both the limbs overturned are

(a) Cross folds (b) Conjugate folds
(c) Tight folds (d) Fan folds

101. The structure having dip towards a common central point from all sides is

(a) Basin (b) Fault
(c) Dome (d) Joint

102. A group of folds having essentially parallel axial planes is

(a) Recumbent folds (b) Isoclinal folds
(c) Conjugate folds (d) Overturned folds

103. Mullions are formed under

(a) Compressive stress regime
(b) Tensile stress regime
(c) Shearing stress regime
(d) Tensile and shearing stress regimes

104. The behavior of perfectly elastic body is governed by

(a) Hook's law (b) Hilt's law
(c) Bulk modulus (d) Bode's law

105. Compressibility can be described as the reciprocal of

(a) Bulk modulus (b) Young's modulus
(c) Rigidity modulus (d) Young's and rigidity modulii

106. Faults striking across structures like fold axes, schistosity, lineation, etc, are known as

(a) Transverse fault (b) Longitudinal fault
(c) Diagonal fault (d) Bedding fault

107. A limited area of younger rocks surrounded by the older rocks is called

(a) Outlier (b) Overlap
(c) Inlier (d) Offlap

108. A group of small size faults overlapping each other in the area of their occurrence is

(a) Parallel faults (b) Radial faults
(c) Encephalon faults (d) Peripheral faults

109. Non-conformity is synonymous to

(a) Heterolithic unconformity
(b) Parallel unconformity

(c) Angular unconformity
(d) Disconformity

110. Gravity faults are formed under:

(a) Compressive stress regime
(b) Tensile stress regime
(c) Shear stress regime
(d) Effects of tensional forces

111. Joints developed perpendicular to the fold axis are termed as:

(a) Columnar joints (b) Release joints
(c) Extension joints (d) Cross joints

112. The greatest principal stress axis is vertical in:

(a) Normal faults (b) Reverse faults
(c) Thrust faults (d) Strike slip faults

113. The distance between any two joints may be described as

(a) Slip cleavage (b) Crenulation cleavage
(c) Fracture cleavage (d) Bedding fassility

114. Schuppen structures are associated with:

(a) Normal faulting (b) Reverse faulting
(c) Thrust faults (d) Recumbent folding

115. Unconformity is which beds are parallel called:

(a) Angular unconformity (b) Disconformity
(c) Non conformity (d) None of these

116. Horst and Grabens are example of:

(a) Fold (b) Unconformity
(c) Fault (d) Joints

117. Part of the Tidal Flats occurring near the high water line is known as:

(a) Mixed flat (b) Sand flat
(c) Mud flat (d) Tidal flat

118. The structure having dip towards a common central point from all sides is

(a) Basin (b) Fault

(c) Dome (d) Joint

119. A group of folds having essentially parallel axial planes are

(a) Recumbent folds (b) Isoclinal folds

(c) Conjugate folds (d) Overturned folds

120. Heterolithic unconformity is also known as

(a) Parallel unconformity (b) Angular unconformity

(c) Disconformity (d) Nonconformity

121. The angle between any line and its horizontal projection, measured in a vertical plane is

(a) Pitch (b) Dip

(c) Plunge (d) Strike

122. The behaviour of perfectly elastic body is governed by

(a) Hilt's Law (b) Bulk Modulus

(c) Hook's Law (d) Bode's Law

123. Compressibility can be described as the reciprocal of

(a) Young's Modulus

(b) Rigidity Modulus

(c) Young's and Rigidity Modulus

(d) Bulk Modulus

124. Joints perpendicular to the axis of fold more common in orogenic belts are termed as

(a) Columnar joints (b) Release joints

(c) Extension joints (d) Cross joints

125. A limited area of younger rocks surrounded by the older rocks is called

(a) Outlier (b) Overlap

(c) Inlier (d) Offlap

126. When two folds plunging away from each other are joined, then they formed

(a) Basin (b) Dome

(c) Culmination (d) Depression

127. Which of the following statements is true?

(a) Pitch and plunge are always same

(b) Fold hinge lines and fold axis are always same

(c) Pitch and rake are always same

(d) None of the above

128. Which of the following is true for passive shear folding?

(a) It develops in single layer competent beds

(b) It develops in single layer incompetent beds

(c) It develops because of bending of the beds

(d) It develops because of micro-faulting across the beds

129. Head of a fault is:

(a) 90° + dip. (b) 90° - dip

(c) Plunge + rake (d) Plunge + dip

130. Which one of the following is true?

(a) The pi –diagram can differentiate between antiform and synform

(b) A vertical fold axis will be plotted at periphery of the stereonet in pi-diagram

(c) An inclined bedding plane will occur as a curve on the pi-diagram

(d) Pole of the bedding plane is plotted in case of pi-diagram

131. Poisson's Ratio is the ratio between:

(a) Young's Modulus and Shear Modulus

(b) Axial strain and transverse strain

(c) Compressibility and rigidity

(d) All of the above

132. When the stream valleys are guided by the folds they are parallel to the system of fold axes, this pattern of drainage is called:

(a) Dendritic (b) Radial

(c) Trellis (d) Annular

133. The smaller folds, which occur on the limbs or hinge of the larger folds, are known as:

(a) Parasitic folds (b) Kink folds
(c) En echelon folds (d) Radial folds

134. A set of new planar surfaces produced in a rock as a result of deformation is called:

(a) Foliation (b) Bedding plane
(c) Lineation (d) Deformation surface

135. If the dip of a plane is measured in a vertical plane perpendicular to the strike, it is called:

(a) True dip (b) Apparent dip
(c) True slope (d) Apparent slope

136. A fault that strikes parallel to the strike of axial plane of the regional fold is known as:

(a) Longitudinal fault (b) Transverse fault
(c) Dip slip fault (d) Diagonal slip fault

137. A major anticline that is composed of many smaller folds is called:

(a) Anticlinorium (b) Synclinorium
(c) Antiformal syncline (d) Synformal anticline

138. Dip of the axial plane and plunge of the fold hinge line are same in a fold. The fold is

(a) Isoclinal (b) Inclined
(c) Reclined (d) Recumbent

139. An upright fold is superimposed by another upright folding with mutually perpendicular fold axes. This will result in a:

(a) Dome and basin structure. (b) Mushroom shaped structure
(c) Hook type structure (d) Positive flower structure

140. A listric fault is

(a) Steep dipping normal fault
(b) Gentle dipping normal fault
(c) Steep dipping fault at top and gental dipping at bottom
(d) Gentle dipping fault at top and steep dipping at bottom

141. Shear joints are formed due to:

(a) Tensional force (b) Compressional force

(c) Coupler force (d) Torsional forces

ANSWER SHEET

1. (b)	2. (d)	3. (d)	4. (c)	5. (a)
6. (d)	7. (c)	8. (a)	9. (d)	10. (d)
11. (c)	12. (a)	13. (c)	14. (d)	15. (c)
16. (b)	17. (c)	18. (b)	19. (b)	20. (c)
21. (a)	22. (b)	23. (c)	24. (b)	25. (d)
26. (b)	27. (b)	28. (a)	29. (d)	30. (a)
31. (c)	32. (a)	33. (d)	34. (c)	35. (b)
36. (c)	37. (c)	38. (c)	39. (a)	40. (d)
41. (d)	42. (a)	43. (c)	44. (d)	45. (b)
46. (b)	47. (c)	48. (d)	49. (c)	50. (b)
51. (b)	52. (b)	53. (c)	54. (b)	55. (a)
56. (b)	57. (c)	58. (a)	59. (c)	60. (d)
61. (b)	62. (a)	63. (c)	64. (a)	65. (b)
66. (b)	67. (a)	68. (c)	69. (c)	70. (b)
71. (c)	72. (c)	73. (b)	74. (a)	75. (d)
76. (c)	77. (b)	78. (c)	79. (a)	80. (b)
81. (c)	82. (b)	83. (b)	84. (b)	85. (a)
86. (b)	87. (c)	88. (c)	89. (a)	90. (a)
91. (b)	92. (a)	93. (a)	94. (d)	95. (c)
96. (b)	97. (b)	98. (b)	99. (d)	100. (d)
101. (a)	102. (b)	103. (a)	104. (a)	105. (a)
106. (c)	107. (a)	108. (c)	109. (a)	110. (b)
111. (a)	112. (a)	113. (c)	114. (c)	115. (b)
116. (c)	117. (c)	118. (a)	119. (b)	120. (d)
121. (c)	122. (c)	123. (d)	124. (c)	125. (a)
126. (c)	127. (c)	128. (d)	129. (c)	130. (d)
131. (b)	132. (c)	133. (a)	134. (a)	135. (a)
136. (a)	137. (a)	138. (c)	139. (a)	140. (c)
141. (a)				

CHAPTER - 6

ORE GEOLOGY

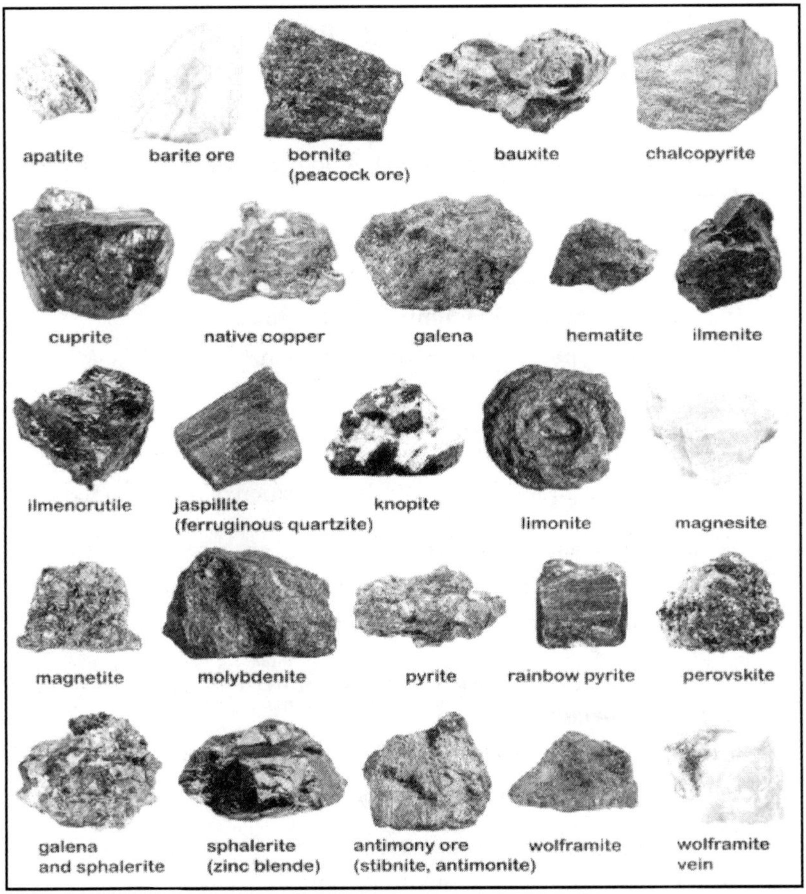

Ore Geology

1. **In which of the following is the highest concentration of 8C13 values found?**
   (a) Black shales (b) Marine Carbonate
   (c) Petroleum (d) Coal

2. **Tin deposit in pegmatite is found in:**
   (a) Ghatsila (b) Mamandur
   (c) Sargipali (d) Jagdalpur

3. **Iron ore orogeny in India is associated with:**
   (a) Achaeans (b) Precambrian
   (c) Palaeozoic (d) Permo- Triassic

4. **Which of the following lists the correct order of coalification process?**
   (a) Anthracite – Bituminous – Lignite – peat
   (b) Bituminous – Lignite – peat – Anthracite
   (c) Peat - Anthracite – Bituminous – Lignite
   (d) Peat – Lignite – Bituminous – Anthracite

5. **Chrysotile asbestos is formed by:**
   (a) Alteration of serpentine (b) Hydrothermal solutions
   (c) Magmatic liquid (d) Alteration of olivine to serpentine

6. **What is Sukinda known for?**
   (a) Mn- ore deposit (b) Chromite deposit
   (c) Fe- ore deposit (d) Base mental deposit

7. **Ag mineralization with minor amount of Cu, Pb and Zn in Rajasthan occurs in:**
   (a) Clay and marlstones (b) Calcareous sandstone
   (c) Siliceous dolomite (d) Schists

8. **Which of the following deposit have strong evidence in support of both hydrothermal and porphyry type of ore deposit?**
   (a) Dariba Rajpura (b) Khetri Deposit
   (c) Zawar Deposit (d) Malanjkhand Deposit

9. Which of the following ore minerals is identified its grey, anhedral and isotropic optical characters?

(a) Cassiterite (b) Hematite
(c) Sphalerite (d) Tennantite

10. Which of the following minerals is identified by high reflectivity?

(a) Pyrite (b) Pyrrhotite
(c) Marcasite (d) Chalcopyrite

11. Which of the following Mn minerals is identified by its well-developed prismatic or tabular habit?

(a) Pyrolusite (b) Psilomelane
(c) Cryotolerance (d) Hollandite

12. Which of the following ores is identified by its coarse gained, euhedral and isotropic characters?

(a) Ilmenite (b) Magnetite
(c) Hematite (d) Goethite

13. Which of the following ore minerals show polygonal cracks filled with gangue?

(a) Cassiterite (b) Manganite
(c) Chromite (d) Braunite

14. Ga is associated with:

(a) Coal (b) Petroleum
(c) Pegmatite (d) Evaporites

15. Chief oil producing horizon in Bombay high is of ______ in age.

(a) Turonian (b) Cenomanian
(c) Albian (d) Maastrichtian

16. Which of the following geological processes has resulted in formation of Precambrian Banded Iron Ore deposits in India?

(a) Sedimentation (b) Hydrothermal activity
(c) Supergene enrichment (d) Magmatic concentration

17. India is an exporter of –

(a) Copper ore (b) Lead-Zink Ore

(c) Iron Ore (d) Rock Phosphate

18. Which of the following is a mis-match?

(a) Gibbsite (b) Diaspore

(c) Halite (d) Bohmite

19. Monazite and Ilmenite deposits of west coast of India are formed by which of the following processes?

(a) Hydrothermal (b) Wind erosion

(c) Residual concentration (d) Magmatic concentration

20. The most significant discovery of gold from Rajasthan in the recent years is from –

(a) South Sindesar (b) Bhukia

(c) Banwas Block (d) Rohil

21. Among following sources of power, India has the largest reserves of:

(a) Coal (b) Oil

(c) Natural gas (d) Atomic power

22. Which of the followings is the use of Deuterium Oxide in nuclear power plants?

(a) As fuel (b) As moderator

(c) A shield (d) As regulator

23. The Lead-Zinc mineralization in Jawar belt is largely confined to which one of the following rocks?

(a) Schist (b) Quartzite

(c) Dolomitic limestone (d) Slate

24. From the chemical point of view smithsonite mineral is -

(a) Zinc carbonate (b) Oxide of Iron and Zinc

(c) Zinc silicate (d) Hydrous Zinc sulphate

25. Geiger-Muller Counter is commonly used for the exploration of -

(a) Bauxite deposits (b) Pb-Zn deposits

(c) Uranium deposits (d) Iron ore deposits

26. Which one of the ore minerals show triangular pits under microscope?

(a) Sphalerite (b) Magnetite

(c) Galena (d) Pyrrhotite

27. Which one is the main constituent of gaseous formation of petroleum?

(a) Methane (b) Ethane

(c) Propane (d) Butane

28. Which of the following is not sedimentary in origin:

(a) Gypsum (b) Halite

(c) Magnetite (d) Hematite

29. Petroleum is generally associated with which rock type:

(a) Igneous (b) Sedimentary

(c) Metamorphic (d) Volcano-sedimentary

30. What is the name of the white clay which has been used for thousands of years in fabrication of ceramic bodies?

(a) Talc (b) Kaolin

(c) Feldspar (d) Quartz

31. Hematite is a primary ore of which of the following metals?

(a) Zinc (b) Tungsten

(c) Iron (d) None of the above

32. Galena is a primary ore of which of the following metals?

(a) Zinc (b) Tungsten

(c) Lead (d) Silver

33. Sphalerite is a primary ore of which of the following metals?

(a) Zinc (b) Tin

(c) Copper (d) Iron

34. Which of the following minerals is a source of copper?

(a) Bauxite (b) Hematite
(c) Malachite (d) Galena

35. The chemical union of iron, oxygen and water forms:

(a) Magnetite (b) Hematite
(c) Limestone (d) Pyrite

36. The oxide of which metal forms corundum, rubies and sapphires?

(a) Aluminium (b) Calcium
(c) Beryllium (d) Potassium

37. Which of the following is the only element which is "mined" from the sea?

(a) Copper (b) Magnesium
(c) Uranium (d) Zinc

38. Which of the following elements is found in the earth's crust as both a native element and as a compound?

(a) Argon (b) Copper
(c) Chlorine (d) Silicon

39. Coal is believed to have been principally formed in which of the following types of environment?

(a) River beds (b) Swamps
(c) Volcanic regions (d) Deserts

40. A geologic feature called a kimberlite indicates the possibility of the presence of which of the following?

(a) Rubies (b) Diamonds
(c) Sapphires (d) Opals

41. Where the oldest and deepest Gold mine in India is located?

(a) Kolar (b) Khetri
(c) Zawar (d) Rampura- Agucha

42. Filament of the electrical bulbs are made from:

(a) Zinc (b) Copper
(c) Tungsten (d) Iron

43. Sanchor basin is situated at:

(a) Bombay (Mumbai) (b) Tamil Nadu
(c) Rajasthan (d) Assam

44. By which process the placer deposit are formed?

(a) Metamorphism (b) Sublimation
(c) Mechanical concentration (d) Sedimentation

45. Which of the following is a fertilizer mineral?

(a) Gypsum (b) Copper
(c) Diamond (d) Feldspar

46. Which of the following is a Gemstone?

(a) Diamond (b) Emerald
(c) Ruby (d) All

47. Chalcopyrite is an ore of

(a) Lead and zinc (b) Copper
(c) Gold (d) Platinum

48. The Zawar is famous for which deposit?

(a) Lead and zinc (b) Copper
(c) Gold (d) Platinum

49. The Jhamarkotra is known for the deposits of?

(a) Rock phosphate (b) Granite
(c) Marble (d) Sandstone

50. Which is the hardest mineral on Earth?

(a) Graphite (b) Corundum
(c) Diamond (d) Muscovite

51. Where in Rajasthan deposits of copper are found?

(a) Rampura-Agucha (b) Khatri
(c) Barmer (d) Degana

52. Which of the following is not an ore of copper?

(a) Hematite (b) Chalcopyrite
(c) Cuprite (d) Malachite

53. Where the supergene sulphide enrichment zone is found?

(a) Above the water table (b) Below the water table

(c) Near the ground surface (d) In oxidizing zone

54. Which is the correct statement about graphite?

(a) It is formed by sedimentary process

(b) It has good conductivity of heat and electricity

(c) It is used in cement industry

(d) All of above

55. The chemical composition of wolframite is -

(a) (Cr, Ni) WO_4

(b) (Fe, Mn)WO_4

(c) Ca WO_4

(d) The interlimb angle is more than 110° hinge is thicker than limbs.

56. The Cryolite mineral was historically used as -

(a) Ore of Chromium (b) Ore of Nickel

(c) Ore of Aluminium (d) Ore of Silver

57. Which state is having the maximum reserves of Talc?

(a) Gujarat (b) Assam

(c) Rajasthan (d) Karnataka

58. Which mineral among the following, indicates the highest temperature and pressure conditions of crystallisation -

(a) Quartz (b) Orthoclase

(c) Hypersthene (d) Biotite

59. The evaporite sequence has which mineral -

(a) Gypsum (b) Coal

(c) Dolomite (d) Stromatolite

60. The major ore mineral, which occur in Rampura-Agucha deposit are -

(a) PbS– ZnS (b) $PbSO_4$ - $ZnSO_4$

(c) NiS - $CrSO_4$ (d) CuS - $CuSO_4$

61. Diamond deposits are associated with which rock?

(a) Lamprophyre (b) Komatiite
(c) Kimberlite (d) Carbonatite

62. The early magmatic segregation process in ultramafic magma develops which of the following?

(a) Wolframite deposit (b) Chromite deposit
(c) Andalusite deposit (d) Wollastonite deposit

63. The late magmatic segregation process is acidic magma develops -

(a) Fluorite deposit (b) Chromite deposit
(c) Talc deposit (d) Asbestos deposit

64. The Giral deposit provider -

(a) China clay (b) Coal
(c) Limestone (d) Mica

65. The Laps Bura mineral deposit is famous for which mineral -

(a) Kyanite (b) Actinolite
(c) Tremolite (d) Graphite

66. The Seismic geophysical survey is done for -

(a) Hydrocarbon Exploration (b) Coal Exploration
(c) Silica Sand Exploration (d) Bauxite Exploration

67. 'Gossan' develops on which type of the deposit

(a) Asbestos (b) Sulphide
(c) Dolomite (d) Coal

68. Diamond drilling method is used for the Exploration of which type of deposit-

(a) Aeolian Deposit (b) Alluvial deposit
(c) Metallic deposit (d) Beach Placer deposit

69. Which of the following represents the correct order of increasing carbon content in different varieties of coal -

(a) Anthracite- bituminous- peat- lignite
(b) Peat- lignite- bituminous- anthracite

(c) Lignite- anthracite- peat- bituminous
(d) Bituminous- peat- lignite- anthracite

70. Which of the following is associated with uranium deposit -

(a) Giral (b) Jaduguda
(c) Sindesar (d) Agnigundala

71. The Rageshwari oil field is situated in which basin -

(a) Cambay Basin (b) Mumbai off shore Basin
(c) Mahanadi Basin (d) Barmer Basin

72. Shahgarh Hydrocarbon block in situated in which state -

(a) Assam (b) W. Bengal
(c) Gujarat (d) Rajasthan

73. Which of the following is not uranium mineral?

(a) Carnotite (b) Uranite
(c) Boehmite (d) Autunite

74. Which variety of coal mined from Panadhro village, Gujarat -

(a) Anthracite (b) Lignite
(c) Peat (d) Bituminous

75. Thorium, extracted from monazite belongs to which one of the following groups?

(a) Oxide (b) Silicate
(c) Phosphate (d) Carbonate

76. Which one of the following is the main Alumina ore mineral in East Coast Bauxite deposit?

(a) Diaspore (b) Gibbsite
(c) Boehrnite (d) Altered Sillimanite

77. Match List-I with List-II and select the correct answer using the codes given below the Lists:

List-I	List-I
(Ore mineral)	(Chemical composition)
(A) Galena	1. PbS
(B) Cerussite	2. $ZnCO_3$

(C) Anglesite | 3. $PbCO_3$

(D) Smithsonite | 4. $PbSO_4$

Code:

	A	B	C	D
(a)	2	4	1	3
(b)	1	4	3	2
(c)	2	3	1	4
(d)	1	3	4	2

78. Which one of the following pairs is not correctly matched?

(a) Hydrothermal : Gold

(b) Magmatic : Chromite

(c) Sedimentation : Kyanite

(d) Residual concentration : Bauxite

79. Match List-I with List-II and select the correct answer using the codes given below the Lists:

List-I (Process)	List-II (Sub process)
(A) Magmatic	1. Cavity filling
(B) Hydrothermal	2. Dissemination
(C) Contact metasomatic	3. Recombination
(D) Residual concentration	4. Weathering

Code:

	A	B	C	D
(a)	4	3	1	2
(b)	2	1	3	4
(c)	4	1	3	2
(d)	2	3	1	4

80. Which types of rocks contain the Graphite deposits of Kalahandi, the Garnet deposits of Rajasthan and the Kyanite deposits of Singhbhum?

(a) Igneous rocks

(b) Sedimentary rocks

(c) Metamorphic rocks
(d) Both Igneous and Sedimentary rocks

81. Which of the following hosts the Chromite deposits of Andhra Pradesh?

(a) Dunite (b) Peridotite
(c) Pyroxenite (d) Amphibolite

82. Match List-I with List-II and select the correct answer using the codes given below the Lists:

List-I (Place)		List-II (Deposits)
(A) Gani-Kalava (Andhra Pradesh)	1.	Chromite
(B) Jamarkotra (Rajasthan)	2.	Copper
(C) Nausahi (Orissa)	3.	Phosphorite
(D) Kollihills (Tamil Nadu)	4.	Bauxite

Code:

	A	B	C	D
(a)	2	3	1	4
(b)	4	1	3	2
(c)	2	1	3	4
(d)	4	3	1	2

83. .... deposits comprise an assemblage of high temperature metamorphic minerals.

(a) Hypothermal (b) Mesothermal
(c) Skarn (d) Residual geo

84. Among the following which one does not belong to the factor affecting the residual concentration?

(a) Climate (b) Local relief
(c) Existence of proper drainage (d) Gravity

85. .... is a hard mantle is produced by useless residual materials and desiccated product of oxidation on the surface.

(a) Gossan (b) Overburden
(c) Gangue (d) Placer

86. Goethite is a/an

(a) Silicate (b) Sulphide
(c) Carbonate (d) Oxide

87. Tenorite is an ore mineral of

(a) Iron (b) Manganese
(c) Copper (d) Lead

88. Which of the following is the ore mineral of manganese?

(a) Smilthsonite (b) Hematite
(c) Bornite (d) Braunite

89. Manganite ore mineral has been found in the rock of:

(a) Gondite (b) Charnockite
(c) Orthoquartzite (d) Marl

90. Bauxite deposits occur as:

(a) Disseminated (b) Loades
(c) Blanket (d) Hydrothermal solution

91. Chromium deposits occur in:

(a) Granite (b) Limestone
(c) Acidic rocks (d) Ultrabasic rocks

92. Kudramukh is famous for:

(a) Banded Hematite Quartz (BHQ) (b) Magnetite
(c) Chromite (d) Philomelone

93. Ruby is a gem variety of

(a) Olivine (b) Labradorite
(c) Quartz (d) Corundum

94. .... is an imperfectly crystallized diamonds with inclusions.

(a) Bort (b) Bullas
(c) Carborado (d) Moonstone

95. Secondary product of Diaspore and Gibbsite is:

(a) Bohemite (b) Bauxite
(c) Hydrogillite (d) Amazon stone

96. Match the following and choose the correct option:

	Column-A		Column-B
P.	Muscovite	1.	Residual Concentration
Q.	Bauxite	2.	Magmatic segregation
R.	Copper	3.	Pegmatite deposits
S.	Chromite	4.	Hydrothermal solution

(a) P-4, Q-2, R-1, S-3
(b) P-3, Q-1, R-2, S-4
(c) P-3, Q-1, R-4, S-2
(d) P-4, Q-2, R-3, S-1

97. Chromate deposits result from:

(a) Early magmatic dissemination
(b) Early magmatic segregation
(c) Early magmatic injection
(d) Late magmatic segregation

98. Ores near Earth's surface are generally obtained from:

(a) Waste removal facilities
(b) Underground mines
(c) Open pit mines
(d) Bodies of water with high concentrations of dissolved minerals

99. In India, the iron ore deposits are mainly of which type?

(a) Magnetite
(b) Hematite
(c) Limonite
(d) Siderite

100. 'Blue Dust' is a type of

(a) Iron deposit in Goa
(b) Copper deposit in Khetri
(c) Manganese deposit in Madhya Predesh
(d) Cobalt deposit

101. Sphalerite is generally associated with

(a) Braunite
(b) Magnetite
(c) Cuprite
(d) Galena

102. Lead and zinc is mainly mined from

(a) Rajasthan
(b) Kerala
(c) Madhya Pradesh
(d) Uttar Pradesh

103. Jaduguda is famous for

(a) Uranium (b) Copper

(c) Manganese (d) Tungsten

104. Kimberlites are important because

(a) They contain crustal Zenoliths

(b) They exhibit fenitisation

(c) They host diamonds

(d) All of the above

105. Which of the following is (are) used in the steel industry?

(a) Coal

(b) Manganese ore

(c) Limestone, dolomite, quartz

(d) All of the above

106. The largest iron are deposits of igneous origin are found in

(a) India (b) Denmark

(c) Sweden (d) Australia

107. Hydrothermal are deposits of mesothermal type are formed at which temperatures?

(a) 0 to 100 ºC (b) 100 to 200 ºC

(c) 200 to 300 ºC (d) 300 to 500 ºC

108. The chromite deposits of Orissa are found in

(a) Sukinda (b) Sitampundi

(c) Ratnagiri (d) Byrapur Area

109. The bauxite deposits of Ranchi have a composition of about:

(a) 20 to 30% Al_2O_3 (b) 40 to 50% Al_2O_3

(c) 50 to 60% Al_2O_3 (d) 10 to 20% Al_2O_3

110. Which mineral is used for the separation of aluminum metal by electrolysis?

(a) Chrysolite (b) Chrysotile

(c) Cryolite (d) Chiastolite

111. Which one of the following can be used to recover gold by amalgamation?

(a) Zine (b) Mercury

(c) Copper (d) Magnesite

112. Which one of the following is a neutral Refractory mineral?

(a) Chromite (b) Kyanite

(c) Sillimanite (d) Magnesite

113. Which one of the following is used in the cement industry?

(a) Zine (b) Pyrite

(c) Gypsum (d) Kyanite

114. Dinosaur egg shells have been found in the Lameta Formation in

(a) Keral (b) Madhya Pradesh

(c) Tamil Nadu (d) Punjab

115. Iron-ore group is known from

(a) Uttar Pradesh (b) Maharashtra

(c) Arunachal (d) Jharkhand-Orissa

116. Chalcopyrite is one mineral of

(a) Aluminium (b) Copper

(c) Iron (d) Silver

117. A vertical dyke showing transverse veins is known as

(a) Stock work (b) Saddle reef

(c) Ladder vein (d) Vug

118. The chemical composition of Hematite is

(a) Fe_2O_3 (b) Fe_3O_4

(c) Fe_3O_2 (d) Fe_2O_4

119. Placer gold deposits are mostly

(a) Elluvial (b) Colluvial

(c) Aeolian (d) Alluvial

120. Banded manganese ores are generally

(a) Epigenetic
(b) Syngenetic
(c) Paragenetic
(d) Both epigenetic and paragenetic

121. Chromite deposits are mostly of

(a) Igneous origin
(b) Metamorphic origin
(c) Sedimentary origin
(d) Both metamorphic and sedimentary origin

122. The chief ore of aluminum is

(a) Pyrolucite (b) Sphalerite
(c) Bauxite (d) Chalcopyrite

123. The most important one of lead is

(a) Rutile (b) Psilomelane
(c) Sphalerite (d) Galena

124. The Kolar Gold Field is located in

(a) Bihar (b) Karnataka
(c) Andhra Pradesh (d) Tamil Nadu

125. 'Smarskite' is an ore mineral of

(a) Thorium (b) Uranium
(c) Cobalt (d) Copper

126. The purest form of iron is

(a) Native iron (b) Pig iron
(c) Wrought iron (d) Steel

127. Uranium deposits or Jaduguda are of

(a) Metamorphic origin (b) Sedimentary origin
(c) Magmatic origin (d) Hydrothermal origin

128. Coal seams are often found to be associated with

(a) China clay (b) Fire clay
(c) Pottery clay (d) Bentonites

129. Plaster of Paris is obtained from

(a) Bauxite (b) Gypsum

(c) Kaolin (d) Limestone

130. The reservoir rock of Bombay High oil field is

(a) Sandstone (b) Limestone

(c) Shale (d) Clay

131. Which of the following is an oxide of titanium and iron?

(a) Rutile (b) Anatase

(c) Hmenite (d) Brookite

132. Streak of Hematite is:

(a) Yellow (b) Green

(c) Orange (d) Cherry red

133. Alumina is ……………..

(a) Hydrated Al_2O_3 (b) $A1_2O_3$

(c) Al metal (d) $Al(OH)_3$

134. Halite is

(a) KCl (b) $CaCl_2$

(c) NaCl (d) $MgCl_2$

135. Placer Gold deposits are mostly

(a) Elluvial (b) Colluvial

(c) Aeolian (d) Alluvial

136. The chief ore of Aluminum is

(a) Pyrolusite (b) Sphalerite

(c) Bauxite (d) Chalcopyrite

137. Diamond is found in region

(a) Panna (b) Kolar

(c) Baster (d) Singhbhum

138. Bornite is an ore of

(a) Iron (b) Lead

(c) Nickel (d) Copper

139. Ajabgarh Group is associated with

(a) Iron (b) Zinc

(c) Copper (d) Manganese

140. Rutile is source for

(a) Tin (b) Titanium

(c) Tungston (d) Iron

141. The most important ore of lead is

(a) Rutile (b) Psilomelane

(c) Sphalerite (d) Galena

142. The Kolar Gold field is located in

(a) Bihar (b) Karnataka

(c) Andhra Pradesh (d) Tamil Nadu

143. 'Smarskite' is an ore mineral of

(a) Thorium (b) Nickel

(c) Cobalt (d) Copper

144. Uranium deposits of Jaduguda are of

(a) Metamorphic origin (b) Sedimentary origin

(c) Magmatic origin (d) Hydrothermal origin

145. Oil in Digboi is found in

(a) Tipam Formation (b) Jaintia Formation

(c) Sylhet Formation (d) Disang Formation

146. Richest bentonite deposit of Gujarat is located in

(a) Jamnagar (b) Mehsana

(c) Kachchh (d) Bharuch

147. Gossan is used to identify a buried:

(a) Sulphide deposit of epigenetic origin

(b) Auriferous vein

(c) Copper porphyry deposit

(d) Multi mineral Zinc- Silver deposit

148. Radon- 222 gas is used in prospecting for:

(a) Uranium (b) Thorium

(c) Silver (d) Antimony

149. The most important hydrochemical indicator of petroleum is:

(a) Naphthenates (b) Ethanol

(c) Humicacids (d) Brines

150. Which one of the following is NOT a manganese mineral?

(a) Pyrolusite (b) Sitaparite

(c) Vredenburgite (d) Smithsonite

151. Which one of the following deposits does NOT occur in Archaean greenstone belts?

(a) Gold (b) Platinum

(c) Aluminum (d) Chromite

152. Which one of the following deposits is formed at oceanic ridges?

(a) Sulphide (b) Carbonate

(c) Oxide (d) Phosphate

153. The Khetri belt of Rajasthan is famous for:

(a) Gold (b) Iron

(c) Copper (d) Coal

154. Which type of tin mineralization occurs in Tosham hills of Haryana?

(a) Cassiterite- Sulphide type

(b) Cassiterite- Quartz and disseminated type

(c) Greisen type

(d) Pegmatite type

155. Which one of the following ore deposits having ferruginous character occurs in Kalahandi area of Odisha?

(a) Bauxite (b) Copper

(c) Galena (d) Diamond

156. Which one of the following minerals is called fool's gold?

(a) Hematite (b) Magnetite
(c) Pyrite (d) Halite

157. The calcareous and siliceous oozes are:

(a) Neritic deposits (b) Littoral deposits
(c) Pelagic deposits (d) Terrigenous deposits

158. The Term 'Petroleum' is derived from the combination of two:

(a) Greek words (b) Latin words
(c) English words (d) Gamow words

159. Out of the following which one is the highest raked coal?

(a) Lignite (b) Anthracite
(c) Bituminous (d) Peat

160. Aluminum is the best substitute for:

(a) Magnesium (b) Nickel
(c) Copper (d) Zine

161. Boghead coals are rich in:

(a) Alginite (b) Sporinite
(c) Fusinite (d) Collotelinite

162. In Seyler's classification, two important parameters considered are:

(a) Carbon and oxygen (b) Hydrogen and oxygen
(c) Carbon and hydrogen (d) Nitrogen and sulphur

163. In coal, the carbon that has not combined with any other element, is regarded as:

(a) Fixed carbon (b) Total elemental carbon
(c) Total organic carbon (d) Inorganic carbon

164. The ratio of 'fixed carbon' to 'volatile matter' contention coal is called as:

(a) Fuel ratio (b) Carbon-hydrogen ratio
(c) Carbon-oxygen ratio (d) Carbon-nitrogen ratio

165. Spheroid or ovoidal, sometimes angular concretions containing mainly carbonates and fossilized plant remains and occur within coal seams are:

(a) Coal ball (b) Ball coal
(c) Eye ball coal (d) Fire ball coal

166. The thermal maturity of coal is determined by measuring reflectance of which of the following:

(a) Inertinite (b) Liptinite
(c) Vitrinite (d) Mineral matter

167. Cannel coals are rich in:

(a) Alginate (b) Sporinite
(c) Suberinite (d) Resinite

168. The optical behaviour of which maceral changes uniformly with increasing rank:

(a) Inertinite (b) Vitrinite
(c) Liptinite (d) Secondary liptinite

169. Coal seams are often found to be associated with:

(a) Fireclays (b) China clays
(c) Bentonites (d) Pottery clay

170. The microlithotype 'durite' is a:

(a) Monomacerite (b) Bimacerite
(c) Trimacerite (d) Tetramacerite

171. Grade of coal is related to:

(a) Carbon content (b) Ash content
(c) Bed moisture (d) Sulphur content

172. Which one of the following is the most effective seal for Petroleum:

(a) Mudstone (b) Shale
(c) Siltstone (d) Evaporite

173. Which one of the following is the most explored petroleum traps?

(a) Structural closure (b) Fault closure
(c) Sub-unconformity traps (d) Salt domes

174. In deep water setting petroleum is often associated with

(a) Organic build ups or "Reefs"

(b) Sub-marine levee channel system

(c) Offshore bars

(d) Marine sabkhas

175. The water saturation in a petroleum reservoir usually:

(a) Increases from top to bottom

(b) Decreases from top to bottom

(c) Increases from east to west

(d) Decreases from north to south

176. The clay used in petroleum refining is:

(a) Kaolin (b) Ball clay

(c) Flint clay (d) Fuller's earth

177. Among following sources of power, India has the largest reserves of:

(a) Coal (b) Oil

(c) Natural gas (d) Atomic power

ANSWER SHEET

1. (b)	2. (d)	3. (b)	4. (d)	5. (d)
6. (b)	7. (c)	8. (d)	9. (c)	10. (b)
11. (d)	12. (b)	13. (c)	14. (a)	15. (d)
16. (d)	17. (c)	18. (c)	19. (d)	20. (c)
21. (a)	22. (b)	23. (c)	24. (a)	25. (c)
26. (c)	27. (d)	28. (c)	29. (b)	30. (b)
31. (c)	32. (c)	33. (a)	34. (c)	35. (c)
36. (a)	37. (c)	38. (b)	39. (b)	40. (b)
41. (a)	42. (c)	43. (c)	44. (c)	45. (a)
46. (d)	47. (b)	48. (a)	49. (a)	50. (c)
51. (b)	52. (a)	53. (b)	54. (b)	55. (b)
56. (c)	57. (c)	58. (c)	59. (c)	60. (a)
61. (c)	62. (b)	63. (a)	64. (b)	65. (a)
66. (a)	67. (b)	68. (c)	69. (b)	70. (b)
71. (d)	72. (d)	73. (c)	74. (b)	75. (c)
76. (b)	77. (d)	78. (a)	79. (b)	80. (c)
81. (b)	82. (a)	83. (c)	84. (d)	85. (a)
86. (d)	87. (d)	88. (d)	89. (a)	90. (c)
91. (d)	92. (b)	93. (d)	94. (a)	95. (b)
96. (c)	97. (b)	98. (c)	99. (b)	100. (a)
101. (d)	102. (a)	103. (a)	104. (c)	105. (a)
106. (c)	107. (c)	108. (a)	109. (c)	110. (c)
111. (b)	112. (a)	113. (c)	114. (b)	115. (d)
116. (b)	117. (c)	118. (a)	119. (d)	120. (b)
121. (a)	122. (c)	123. (d)	124. (b)	125. (a)
126. (c)	127. (d)	128. (b)	129. (b)	130. (b)
131. (c)	132. (d)	133. (b)	134. (c)	135. (d)
136. (c)	137. (a)	138. (d)	139. (c)	140. (b)
141. (d)	142. (b)	143. (a)	144. (d)	145. (a)
146. (c)	147. (a)	148. (a)	149. (a)	150. (d)
151. (c)	152. (a)	153. (c)	154. (c)	155. (a)
156. (c)	157. (c)	158. (b)	159. (b)	160. (a)
161. (a)	162. (c)	163. (a)	164. (a)	165. (a)
166. (c)	167. (b)	168. (b)	169. (a)	170. (b)
171. (b)	172. (d)	173. (a)	174. (b)	175. (a)
176. (d)	177. (a)			

CHAPTER - 7

PALAEONTOLOGY

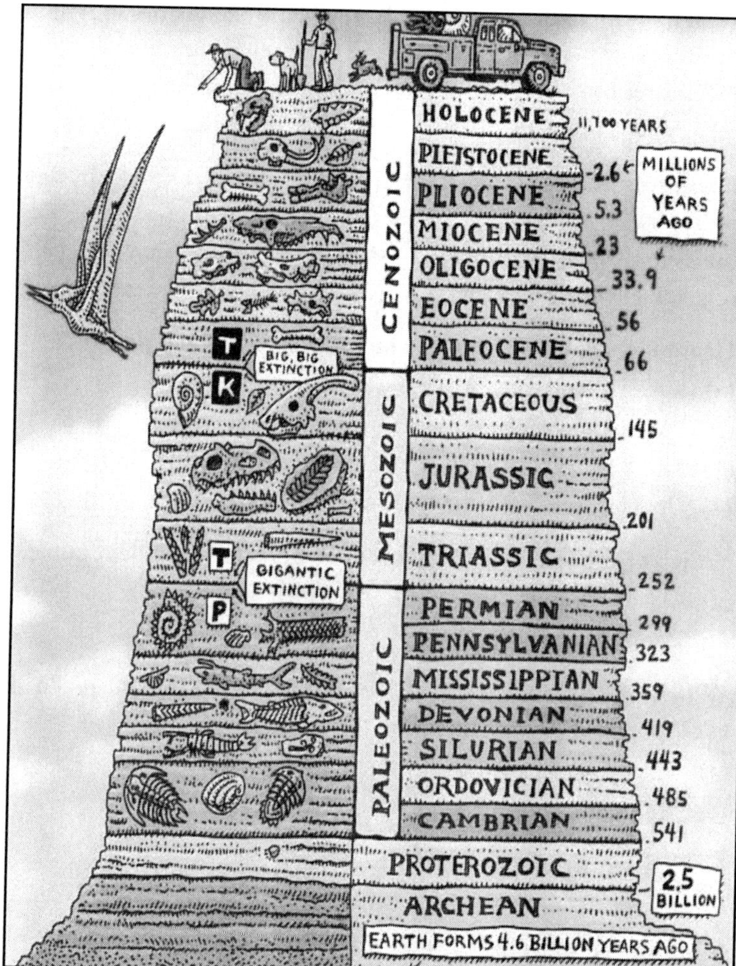

PALAEONTOLOGY

1. **Which of the best index fossil for permo-carboniferous rocks?**
   (a) Products (b) Spirifer
   (c) Syringothyris (d) Lingula

2. **In which of the following periods were Echinoids rarely found?**
   (a) Late Permian (b) Triassic
   (c) Jurassic (d) Paleocene

3. **Which of the following period is marked by index fossil Trilobites Microdiscus?**
   (a) Lower Cambrian (b) Middle Cambrian
   (c) Upper Cambrian (d) None of theses

4. **Fossils found in the stratigraphic succession of Daltonganj clearly indicate:**
   (a) River water intrusion (b) Sea water intrusion
   (c) Pelagic environment (d) Littoral environment

5. **Graptolites are confined to which of the following periods?**
   (a) Lower Cambrian – Middle Cambrian
   (b) Lower Cambrian – Lower Carboniferous
   (c) Middle Cambrian – Lower Carboniferous
   (d) Upper Cambrian – Lower Carboniferous

6. **Which of the following fauna represents bifurcated leaf margins?**
   (a) Gondwanidium (b) Eukaryota
   (c) Barakardichloma (d) Ptilophyllumaccutifolia

7. **Which is the earliest fossil remains of the horse in equine evolution?**
   (a) Orohippus (b) Epihippus
   (c) Mohippus (d) Mesohippus

8. **Land plant first appeared in:**
   (a) Cambrian (b) Ordovician
   (c) Silurian (d) Devonian

9. Which of the following is an example of Diamorphism?

(a) Nummulites (b) Discocyclina

(c) Textularia (d) Rotalia

10. Which of the following is not a suitable method for dating Precambrian rocks?

(a) 14C dating (b) Rb-Sr dating

(c) Sm-Nd dating (d) U-Pb dating

11. Which of the following is a compound coral?

(a) Calceola (b) Zaphrentis

(c) Montlivaltia (d) Isastrea

12. Ediacaran Fossils are useful in determining ....

(a) Archaean/Proterozoic boundary

(b) Precambrian/Cambrian boundary

(c) Permian/Triassic boundary

(d) Cretaceous-Tertiary boundary

13. Which of the following epochs are included in Neogene period?

(a) Palaeocene and Eocene

(b) Palaeocene - Eocene and Oligocene

(c) Oligocene - Miocene as Pliocene

(d) Miocene and Pliocene

14. The original fossil from which a species is formally describe is called -

(a) Paratype (b) Holotype

(c) Archetype (d) Lactotype

15. Biplication feature exhibit in animals belonging which one of the followings?

(a) Brachiopoda (b) Gastropoda

(c) Bivalvia (d) Cephalopoda

16. Which one of the following type of Coral reefs, Lakshadweep reef are?

(a) Fringing reef (b) Barrier reef

(c) Atolls (d) Both (1) and (2)

17. In pelecypoda shell, if umbo is curve in such a way that the beak points in an anterior direction, the umbo are -

(a) Opisthogyrate (b) Prosogyrate
(c) Orthogyrate (d) Spurogyrate

18. Which one of the following animals us consider as largest terrestrial carnivorous that has ever lived on the Earth?

(a) Spinosaurus (b) Brontosaurus
(c) Tyrannosaurus (d) Ornithotestes

19. Which form of Gastropods has spire that is almost concealed by the last whorl?

(a) Fusinus (b) Cypraea
(c) Trochus (d) Conus

20. Which formation of Siwalik Supergroup is richest in vertibrate fossils?

(a) Kamlial (b) Dhok Pathan
(c) Nagri (d) Pinjor

21. Which one of the following is a typical lower Gondwana assemblage?

(a) Glossopteris, Ptillophyllum, Nilsonia, Bucklandia
(b) Glossopteris, Gangmopteris, Brachyphyllum, Nilsonia
(c) Gangmopteris, Lycopodites, Brachyphyllum, Nilsonia
(d) Vertebraria, Alethopteris, Otozamite, Glossopteris

22. Which of the following are restricted to Paleozoic rocks only ?

(a) Bivalvia (b) Cephalopoda
(c) Echinoidea (d) Graptolithina

23. The shallow shelf environment to dominate by

(a) Molluscs and brachiopod lingula
(b) Trace fossils such as Nereites
(c) Pelagic organisms
(d) Bryozoans and branchiopods

24. Foramina are the holes

(a) In the wall of the animals's test
(b) In the septa that allows communication between Chambers

(c) In the marginal cord that forms spiral lamina in Nummulites

(d) That forms the aperture in the test

25. 'Nummulitecs' is a team to indicate shallow neritic deposits rich in large Foraminifers, especially Nummulitecs'. Which of the following period was dominated by them ?

(a) Piocene (b) Eocene

(c) Permian (d) Triassic

26. Fossils are abundantly found in

(a) Igneous rocks (b) Sedimentary rocks

(c) Metamorphic to rocks (d) Evaporites

27. Olduvai-magnetic event is marked by the first appearance of mean on the globe and is represented in:

(a) Gilbert Epoch (b) Matuyama Epoch

(c) Gauss Epoch (d) Brunches Epoch

28. Process of petrifaction results in the formation of-

(a) Fossils (b) Sediments

(c) Petroleum (d) Gas hyderate

29. In which of the following branch of geology are fossils useful?

(a) Landslide zonation (b) Study of palaeoenvironment

(c) Petrogenesis (d) Metamorphism

30. Which of the following mineral substance is used by molluscan animals and is commonly found in the form of fossils?

(a) Calcium carbonate (b) Chitin

(c) Silica (d) Calcium phosphate

31. Oldest fossils occur in the rocks of which age?

(a) Cambrian (b) Archaean

(c) Vendian (d) Tommotian

32. Dinosaur bones are most commonly fossilised as ...

(a) Compressed remains (b) Moulds

(c) Permineralised remains (d) Gastroliths

33. The science dealing with the order of deposition of sedimentary rock based on fossils is described as:

(a) Biostratigraphy (b) Binomial classification

(c) Geochronology (d) Lithostratigraphy

34. Trilobites are so-called because...

(a) Their body can be divided into three transverse divisions called cephalon, thorax and pygidium

(b) They were first studied by an Australian palaeontologist who named them Trilobites

(c) Their body can be divided into three longitudinal divisions

(d) These are living fossils

35. For establishing the precise biozonation of a fossiliferous outcrop sequence, which kind of fossil data is used?

(a) Assemblage of fossils

(b) First appearance of a fossil taxon

(c) Abundant data of fossils

(d) Abundant assemblage of the fossils

36. One of the following, the oldest Mesozoic is-

(a) Triassic (b) Silurian

(c) Jurassic (d) Eocene

37. The characteristics feature of the Mesozoic Era has been-

(a) Fragmentation of Gondwanaland

(b) The dominance of reptiles

(c) Relatively high temperature

(d) Appearance of man

38. Which of the following gives the correct sequence from older to younger ?

(a) Cambrian-Devonian-Permian-Carboniferous- Cretaceous

(b) Cambrian-Ordovician-Permian-Carboniferous- Cretaceous

(c) Cambrian-Silurian-Carboniferous-Triassic- Cretaceous

(d) Cambrian-Devonian-Permian-Carboniferous- Cambrian

39. The range of the tertiary period is

(a) 65-18 Ma
(b) 150-100 Ma
(c) Less than 10 Ka
(d) 150-125 Ma

40. The older rocks that contain abundant fossils are from:

(a) Precambrian era
(b) Paleozoic era
(c) Mesozoic era
(d) Cenozoic era

41. The longest era was the

(a) Mesozoic
(b) Precambrian
(c) Cenozoic
(d) Paleozoic

42. Fossils are rarely in such rocks as:

(a) Limestone
(b) Sandstone
(c) Shale
(d) Granite

43. A distinctive fossil that exists over a relatively large part is earth during a shirt time of earth history

(a) Index fossil
(b) Trance fossil
(c) Universal fossil
(d) True fossil

44. In an ocean, the upper layer of water with sufficient light penetration to support photosynthesis is by aquatic plants is known as the:

(a) Free zone
(b) Plant zone
(c) Photic zone
(d) Primary zone

45. An open-ocean community is called a :

(a) Pelagic community
(b) Benthic community
(c) Lentic community
(d) Lotic community

46. What are fossils?

(a) Organisms buried in the recent sediments
(b) Organisms which lived in the geological past
(c) Remains of organisms lived in the geological past and buried under the sediments
(d) Somcthing dug up

47. Why ammonites are important for biostratigraphers?

(a) They are well preserved
(b) They are index fossils
(c) They are extinct group of animals
(d) They are marine animals

48. Why fossils are useful

(a) They are decorative pieces
(b) They occur in Rajasthan
(c) They give many information about the geological past
(d) None of the above

49. What is a Glossopteris?

(a) It is a leaf-fossil of lower Gondwana
(b) It is an index fossil of Gondwana super group
(c) It is a leaf fossil of Cretaceous
(d) It is a fossil-wood

50. What is physa?

(a) It is a leaf-fossil
(b) It is a dextrally coiled gastropod
(c) It is a sinistrally coiled gastropod
(d) None of the above

51. Calymene is a trilobite

(a) With spines (b) Without spines
(c) With very large eyes (d) Without eyes

52. Productus is a

(a) Branchiopod (b) Lamellibranche
(c) Gastropod (d) Cephalopod

53. Lamellibranches are generally

(a) Inequilateral and equivalve
(b) Equilateral and inequivalve
(c) Single valved
(d) Of very small size

54. The age of Hippurites is

(a) Paleozoic (b) Jurassic

(c) Cretaceous (d) Recent

55. What are Foraminifers ?

(a) Nano-organisms (b) Macro-organisms

(c) Micro-organisms (d) None of the above

56. Ichnology deals with?

(a) Trace-fossils (b) Plant fossils

(c) Micro fossils (d) Body fossils

57. Regular echinoids have:

(a) Always circular test (b) Occasionally circular test

(c) Never circular test (d) No test

58. What are Graptolites?

(a) Small organisms

(b) Fossil which look as if something is written on the stone

(c) Black coloured fossils

(d) None of these

59. Brachial skeleton is found in

(a) Lamellibranchs (b) Cephalopods

(c) Brachiopods (d) Gastropods

60. Invertebrate palaeontology deals with

(a) All the fossils

(b) Only vertebrate animals fossils

(c) Only invertebrate animals fossils

(d) Plant fossils

61. Murex is a

(a) Plant fossil (b) Vertebrate animal

(c) Invertebrate animal (d) Micro organisms

62. Vertical distribution of fossils and geological time-scale are closely related it is :

(a) Not true (b) True

(c) Partially true (d) None of the above

63. Trilobites are found in:

(a) Precambrian rocks (b) Palaeozoic rocks

(c) Mesozoic rocks (d) Cenozoic rocks

64. Belemnites belong to which of the following:

(a) Nautiloda (b) Ammonoidea

(c) Dibranchiate (d) None of the above

65. Certain clay layers deposited at the end of the Cretaceous period are enriched in what element?

(a) Aluminum (b) Carbon

(c) Iridium (d) Uranium

66. Which geologic period not belongs in the Mesozoic era?

(a) Permian (b) Cretaceous

(c) Triassic (d) Jurassic

67. Which of the following is the oldest period of time lasting from the beginning of the earth to approximately 570 million years ago?

(a) Precambrian (b) Pleistocene

(c) Miocene (d) Jurassic

68. Which of the following time period include part of the geologic range of dinosaurs?

(a) Mississippian (b) Devonian

(c) Jurassic (d) None of the above

69. On the phanerozoic time scale, which of the following geologic periods occurred approximately 65 to 136 million years ago? Is it:

(a) Cambrian (b) Ordovician

(c) Carboniferous (d) Cretaceous

70. What is the term used to denote fossil fecal matter -

(a) Moulds (b) Casts

(c) Coprolites (d) Track and trail's

71. The study of fossil pollen grains and spores is known as -

(a) Ethnology (b) Palynology

(c) Palaeo-ecology (d) Pedology

72. Which of the following represents the earliest stage in the evolution of the Horse?

(a) Eohippus (b) Merychippus

(c) Pliohippus (d) Equs

73. Calymene fossil belongs to which of the following:

(a) Brachiopoda (b) Trilobita

(c) Lamellibranchia (d) Echinodia

74. The Graptoloidea on the Earth existed during which time -

(a) Palaeozoic (b) Mesozoic

(c) Neo Proterozoic (d) Meso Proterozoic

75. Which one of the following statements correctly defines the term 'swimmer'?

(a) Plankton (b) Nekton

(c) Benthos (d) Bottom creeper

76. Which one of the following is the principle and essential condition to be an index fossil?

(a) Wide horizontal distribution

(b) Short vertical distribution

(c) Peculiar morphological feature

(d) More amount of fossils

77. Which one of the following is monomyaria and opisthocline type of bivalvia?

(a) Pecton (b) Cardita

(c) Lucina (d) Exogyra

78. Which one of the following is not biconvex?

(a) Spirifer (b) Orthis

(c) Productus (d) Terebratula

79. Which one of the following Gasteropods is planispirally coiled?

(a) Murex (b) Turritella

(c) Bellerophon (d) Pleurotomaria

80. Which one of the following represents an uncoiled form of cephalopoda?

(a) Nautilus (b) Orthocerus

(c) Goniatite (d) Ceratites

81. Which one of the following is regular Echinoid?

(a) Hemiaster (b) Micraster

(c) Cidaris (d) Echinolampus

82. Match List-I with List-II and select the correct answer using the codes given below the Lists:

	List-I		List-II
A.	Pedicle opening	1.	Trilobita
B.	Thorax	2.	Echinoidea
C.	Periproct	3.	Brachiopoda
D.	Cardinal fossils	4.	Anthozoa

Code:

	A	B	C	D
(a)	3	2	1	4
(b)	4	2	1	3
(c)	3	1	2	4
(d)	4	1	2	3

83. What is correct age of Inoceramus?

(a) Jurassic to Cretaceous (b) Triassic to Eocene

(c) Triassic to Recent (d) Cretaceous to Recent

84. What is the common distribution range of Ammonoids?

(a) Jurassic to Tertiary (b) Triassic to Cretaceous
(c) Permian to Jurassic (d) Triassic to Jurassic

85. Which period marks the disappearance of the Rugose corals?

(a) Triassic period (b) Jurassic period
(c) Permian period (d) Cretaceous period

86. Which one of the following is a cow amongst Siwalik mammals?

(a) Bos (b) Sus
(c) Suidae (d) Equus

87. Which one of the following fossils is the odd one?

(a) Ptillophyllum (b) Otozamites
(c) Williamsonia (d) Glossopteris

88. The Paleozoic are is restricted between

(a) 65 to 225 Ma (b) 1.8 to 65 Ma
(c) 570 to 2500 Ma (d) 225 to 570 Ma

89. Which era is called as 'Age of Reptiles'

(a) Paleozoic era (b) Mesozoic era
(c) Ceinozoic era (d) Proterozoic era

90. Which is not an invertebrate fossil in the following?

(a) Productus (b) Ptylophylum
(c) Trigonia (d) Murex

91. Which one is an index fossil?

(a) Trilobite (b) Natiloids
(c) Gastropods (d) Bivalves

92. Match the following:

Column-A (Phylum)	**Column-B (Species)**
P. Brachiopoda	1. Turritella
Q. Cephalopoda	2. Goniatite
R. Gastropoda	3. Sprifer
S. Bivalvia	4. Spondylus

(a) P-1, Q-2, R-3, S-4 (b) P-1, Q-3, R-2, S-4
(c) P-3, Q-1, R-4, S-2 (d) P-3, Q-2, R-1, S-4

93. What are the characteristics of an index fossil?

(a) Wide geographical range and limited time range
(b) Both time and geographical ranges should be wide
(c) Limited geographical range and wide time range
(d) None of the above

94. Trigonia has ....dentition pattern

(a) Isodont (b) Schizodont
(c) Heterodont (d) Desmodont

95. The coiling of shell in ....is sinistral

(a) Trochus (b) Murex
(c) Planoobis (d) Physa

96. Which is not the sub class of Cephalopod a?

(a) Nautiloidea (b) Ammonoidea
(c) Acephala (d) Dibranchia

97. The body of trilobite is made up of three distinct parts known as cephalic shield, thorax and ....

(a) Ocular plates (b) Genital plates
(c) Corona (d) Pygidium

98. Ammonoidea is a/an

(a) Trace fossil (b) Index fossil
(c) Living fossil (d) Plant fossil

99. Ceratitic suture lines have

(a) Pointed lobes and rounded saddle
(b) Pointed lobes as well as rounded saddle
(c) Divided lobes and rounded saddle
(d) Divided lobes and divided saddles

100. Mouth and anus are centrally located in

(a) Regular Echinoids
(b) Irregular echinoids

(c) Both regular and irregular echinoids
(d) Neither regular echinoids nor irregular echinoids

101. Both the values are held together by means of muscles in

(a) Articulate brachiopods
(b) Inarticulate brachiopods
(c) Both (1) and (2)
(d) None of the above

102. Bar all series overlain by

(a) Tipam series
(b) Dupitila series
(c) Surma series
(d) Jaintia series

103. The shell is equilateral and unequal in

(a) Acephala
(b) Brachiopoda
(c) Gastropoda
(d) Ammonoidea

104. Pupilla is a:

(a) Discoidal shape gastropoda
(b) Conical shape gastropoda
(c) Cylindrical shape gastropoda
(d) Globular shape gastropoda

105. Mamelon is the part of ....of the Echinoids.

(a) Apical disc
(b) Peristome
(c) Spines
(d) Tubercle

106. Trilobites become extinct at the end of:

(a) Cambrian
(b) Devonian
(c) Permian
(d) Cretaceous

107. Graptolites reached their maximum development in:

(a) Cambrian
(b) Ordovician
(c) Silurian
(d) Carboniferous

108. Ammonoids become extinct at the end of:

(a) Tertiary
(b) Triassic
(c) Cretaceous
(d) Carboniferous

109. Foramen is found in which one of the following?

(a) Trilobites (b) Pelecypods

(c) Graptolites (d) Brachiopods

110. Which one of the following is a coral?

(a) Matlivaltia (b) Globigerina

(c) Cardita (d) Calymene

111. Which one is a plant fossil?

(a) Schizoneura (b) Favosties

(c) Productus (d) Syringothyris

112. Ichnofossil is

(a) An index fossil (b) A trace fossil

(c) A living fossil (d) None of these

113. The base of Palaeozoic Era is marked by first appearance of

(a) Corals (b) Trilobites

(c) Brachiopods (d) Cephalopods

114. Triasic begins with first appearance of:

(a) Olenus (b) Nautilus

(c) Otoceraswoodwardi (d) Macrocephalites

115. The close of Cretaceous marks the extinction of:

(a) Bivalves (b) Trilobites

(c) Corals (d) Dinosaurs

116. The tooth like phosphatic microfossils are:

(a) Conodonts (b) Thecodonts

(c) Bathydonts (d) Coprolites

117. Cephalopodan with complex suture is:

(a) Ceralites (b) Nautilus

(c) Goniatites (d) Ammonites

118. The trilobite with pygidium equal to that of cephalon is called:

(a) Micropygous (b) Macropygous

(c) Isopygous (d) Either micropygous or isopygous

119. Which one of the following trilobites is characteristic of Middle Cambrian?

(a) Redlichia (b) Olenellus
(c) Paradoxides (d) Olenus

120. The order Primates is for:

(a) Homo sapiens (b) Trilobites
(c) Brachiopods (d) Mollusca

121. The gastropod genus Physa is:

(a) Uncoiled (b) Dextrally coiled
(c) Sinistrally coiled (d) Planispirally coiled

122. The trilobite having large number of lenses in the eyes is:

(a) Agnostus (b) Microdicus
(c) Remopleurides (d) Olenellus

123. The exoskeleton of mollusks is most commonly composed of:

(a) Silica (b) Calcite
(c) Aragonite (d) Phosphate

124. The least favourable environment for the preservation of fossils is:

(a) Terrestrial (b) Lacustrine
(c) Fluvial (d) Marine

125. Which one is not a bivalve?

(a) Nautilus (b) Lima
(c) Nucula (d) Trigonia

126. Rapidly evolving class is:

(a) Bivalvia (b) Gastropoda
(c) Cephalopoda (d) Brachiopoda

127. Trilobites got extinct towards the close of:

(a) Devonian (b) Permian
(c) Cambrian (d) Silurian

128. The characteristic flora of Middle Gondwana is:

(a) Glossoptens (b) Ptiophyllum
(c) Dichroidium (d) Gangamopteris

129. Katorl formation belongs to:

(a) Triassic of Spiti (b) Jurassic of Jaisalmer
(c) Jurassic of Kachchh (d) Cretaceous of Jabalpur

130. In brachiopods, the pedical valve is also called as:

(a) Brachial valve (b) Ventral valve
(c) Right valve (d) Dorsal valve

131. The corona is the part of ....Shell

(a) Ammonoid (b) Trilobite
(c) Echinoid (d) Bivalve

132. The post-mortem history of fossils is studied under the branch of:

(a) Ichnology (b) Palaeoecology
(c) Biostratigraphy (d) Taphonomy

133. In brachiopods, the pedical valve is also called as:

(a) Brachial valve (b) Ventral valve
(c) Right valve (d) Dorsal valve

134. Which of the following is a sinistrally coiled Gastropod?

(a) Physa (b) Murex
(c) Turbo (d) Natica

135. In echinoids, Aristotle's Lantern is found in:

(a) Periproct (b) Peristome
(c) Apical Disc (d) Along lateral margins

136. Which of the following is a characteristic of the Upper Gondwana plant fossil?

(a) Glossopteris (b) Pecopteris
(b) Dicrodium (d) Ptilophyllum

137. Which of the following marks the Permian/Triassic boundary?

(a) First appearance of Olenus
(b) First appearance of Nautilus
(c) First appearance of Macrocephalites
(d) First appearance of Otoceraswoodwardi

138. Cephalopoda with most complex suture is:
(a) Ceratites (b) Ammonites
(c) Nautilus (d) Goniatite

139. The exoskeleton of Pteropods is made up of:
(a) Aragonite (b) Calcite
(c) Calcium phosphate (d) Silica

140. Homo sapiens belong to:
(a) Primates (b) Rodentia
(c) Chiroptera (d) Mollusca

141. Which one is not a bivalve?
(a) Lima (b) Nautilus
(c) Nucula (d) Trigonia

142. Ediacaran fauna belong to:
(a) Mesoproterozoic (b) Palaeoproterozoic
(c) Neoproterozoic (d) Archaeozoic

143. Composition of the diatom microfossil tests is:
(a) Calcareous (b) Siliceous
(c) Phosphatic (d) Cutinous

144. Foraminifera belong to the Kingdom:
(a) Protista (b) Monera
(c) Planta (d) Animalia

145. The sister branch of Paleontology, which deals with the study of Pollens and Spores, is called:
(a) Ichnology (b) Paleobotany
(c) Palynology (d) Oncology

146. The Coccoliths are microscopic:
(a) Marine algae (b) Non-marine algae
(c) Marine fungi (d) Non-marine fungi

147. A depth in ocean below which no calcareous tests are preserved in sediments is called:
(a) Aragonite Compensation Depth

(b) Calcite Compensation Depth
(c) Carbonate lysocline
(d) Supra-lysocline

148. The microfossil group Ostracoda belongs to the phylum:
(a) Mollusca (b) Arthropoda
(c) Bryozoa (d) Protozoa

149. Which of the following microfossil groups is useful in stratigraphic correlation of a marine with a non-marine sequence?
(a) Foraminifera (b) Radiolaria
(c) Pteropoda (d) Ostracoda

150. Abundance of planktonic foraminifera from shallow shelf towards deep slope region:
(a) Increases (b) Decreases
(c) Remains same (d) No definite pattern

151. Occurrence of a larger foraminifera Fusulina in a sedimentary formation suggests an age of:
(a) Permian (b) Carboniferous
(c) Triassic (d) Jurassic

152. Consider the following statements regarding Gastropods:
1. They are typically benthi, though pelagic forms do occur
2. Forms with a siphonal canal are generally carnivorous
3. Forms in which the shell has an entire aperture are often herbivorous
4. Fresh water gastropods for the most part have thin shells with a thick periostracum. Which of the statements given above are correct?

(a) 1, 2 and 4 only (b) 1 and 4 only
(c) 1, 2, 3 and 4 (d) 2, 3 and 4 only

153. Ammonoids belong to which one of the following geological ranges?
(a) Cambrian–Jurassic (b) Ordovician–Triassic
(c) Silurian–Jurassic (d) Devonian–Cretaceous

154. The entire skeleton of a solitary or of a colonial coral is known as:

(a) Corallum (b) Corallite
(c) Columella (d) Coenenchyme

155. The triangular gap along the hinge line of the pedicle valve of brachiopods, through which the pedicle emerges, is known as:

(a) Deltidium (b) Delthyrium
(c) Brachidium (d) Brachiophore

156. The moulting of the exoskeleton in arthropods is known as:

(a) Ecdysis (b) Metamorphosis
(c) Integument (d) Exuvia

157. Which one of the following is a phosphatic microfossil?

(a) Foraminifera (b) Radiolaria
(c) Cocolithophore (d) Conodonts

158. Which one of the following is NOT a seed fern?

(a) Pecopteris (b) Neuropteris
(c) Alethopteris (d) All of these

159. Taxodont dentition is present in:

(a) Arca (b) Venus
(c) Mya (d) Trigonia

160. Cephalopods having complex suture are:

(a) Ceratites (b) Nautilus
(c) Goniatites (d) Ammonites

161. The study of interrelationships between the ancient organisms and the environment in which they lived, is done under:

(a) Palaeoecology (b) Biostratigraphy
(c) Ichnology (d) Taphonomy

162. The marginal or ventral facial suture in trilobites is known as:

(a) Prpparian (b) Opisthoparian
(c) Hypoparian (d) Gonatoparian

163. The amphidetic ligament, whose maximum length is transverse to the commissure of a bivalve is called:

(a) Multivincular (b) Parvincular

(c) Alivincular (d) Duplivincular

164. Select an oldest bivalve genus in the geological record from the following:

(a) Nucula (b) Megalidon

(c) Eurydesma (d) Fordilla

165. Which one of the following is an example of living fossil?

(a) Tetebratula (b) Rhynchonella

(c) Lingula (d) Hippurites

166. The bryozoans colony having sheet like appearance and with zooids on both sides, is known as:

(a) Foliaceous (b) Fenestrate

(c) Dendroid (d) Encrusting

167. The valves in branchiopods are opened by:

(a) Adductor muscles (b) Diductor muscles

(c) Ligaments (d) Pallial sinus

168. Which one of the following microfossils has a siliceous shell?

(a) Ostracodes (b) Diatoms

(c) Pteropods (d) Cocoliths

169. Wandering animals on sea-bottom are known as

(a) Sessile (b) Nektic

(c) Vagile (d) Planktic

170. Who wrote the book systema natures?

(a) Carl von Linnaeus (b) Charles Darwin

(c) William Smith (d) Charles lyell

171. A species, in linnaeus concept is called?

(a) Biospecies (b) Morphospecies

(c) Chronospecies (d) Evolutionary species

172. A basic mechanism of growth, the periodic shedding of exoskeleton and formation of a new one, in trilobites, is known as:

(a) Modification (b) Ecdysis

(c) Accretion (d) Anisometric growth

173. An endogastriccephalopoda is:

(a) Orthoconic in shape

(b) Cyrtoconic, with siphuncle on the concave side of the shell

(c) Cyrtoconic, with siphuncle on the ventral side of the shell

(d) Breviconic in shape

174. Which one of the following is not a typical intertrappean plant?

(a) Palmoxylon (b) Nipadites

(c) Dicroidium (d) Azolla

175. The fundamental unit of chronostratigraphic classification is:

(a) Erathem (b) System

(c) Stage (d) Series

176. Find odd one out

(a) Period (b) Member

(c) Epoch (d) Age

ANSWER SHEET

1. (b)	2. (c)	3. (a)	4. (b)	5. (d)
6. (d)	7. (d)	8. (c)	9. (a)	10. (d)
11. (a)	12. (b)	13. (d)	14. (b)	15. (c)
16. (a)	17. (b)	18. (a)	19. (b)	20. (d)
21. (a)	22. (d)	23. (a)	24. (b)	25. (b)
26. (b)	27. (b)	28. (a)	29. (b)	30. (a)
31. (b)	32. (c)	33. (a)	34. (c)	35. (b)
36. (a)	37. (d)	38. (c)	39. (a)	40. (b)
41. (b)	42. (d)	43. (a)	44. (c)	45. (a)
46. (c)	47. (b)	48. (c)	49. (a)	50. (c)
51. (b)	52. (a)	53. (a)	54. (c)	55. (c)
56. (a)	57. (a)	58. (b)	59. (c)	60. (c)
61. (c)	62. (c)	63. (b)	64. (b)	65. (c)
66. (a)	67. (a)	68. (c)	69. (d)	70. (c)
71. (b)	72. (a)	73. (b)	74. (a)	75. (d)
76. (a)	77. (d)	78. (b)	79. (b)	80. (b)
81. (c)	82. (d)	83. (a)	84. (c)	85. (c)
86. (b)	87. (d)	88. (d)	89. (b)	90. (b)
91. (a)	92. (d)	93. (a)	94. (b)	95. (d)
96. (c)	97. (d)	98. (b)	99. (c)	100. (a)
101. (c)	102. (b)	103. (b)	104. (c)	105. (c)
106. (c)	107. (b)	108. (d)	109. (d)	110. (a)
111. (a)	112. (b)	113. (b)	114. (c)	115. (d)
116. (a)	117. (d)	118. (c)	119. (c)	120. (a)
121. (c)	122. (c)	123. (b)	124. (a)	125. (a)
126. (c)	127. (b)	128. (c)	129. (c)	130. (b)
131. (c)	132. (d)	133. (b)	134. (a)	135. (b)
136. (d)	137. (d)	138. (b)	139. (a)	140. (a)
141. (b)	142. (c)	143. (b)	144. (a)	145. (c)
146. (a)	147. (b)	148. (b)	149. (d)	150. (a)
151. (a)	152. (c)	153. (d)	154. (a)	155. (b)
156. (a)	157. (d)	158. (d)	159. (a)	160. (d)
161. (a)	162. (c)	163. (c)	164. (d)	165. (c)
166. (a)	167. (b)	168. (b)	169. (c)	170. (a)
171. (a)	172. (b)	173. (b)	174. (c)	175. (c)
176. (b)				

CHAPTER - 8

STRATIGRAPHY

STRATIGRAPHY

1. **Greenstone belt is associated with:**
   (a) BGC (b) Bastar Crayons
   (c) Sargoors (d) Mahakaoshals

2. **Mn – bearing horizon in Central India is associated with which of the following stratigraphic successions?**
   (a) Saussars (b) Sakolis
   (c) Chilpis (d) Bijawars

3. **Late Maastrichtian fresh water facies of sediments in India are called:**
   (a) Motur sandstone beds (b) Bagh beds
   (c) Lameta beds (d) Baripada beds

4. **Which of the following is equivalent to Gwalior system?**
   (a) Lower Vindhyan system (b) Cuddapah system
   (c) Dharwar system (d) Aravallis

5. **Sole marks having longitudinal appearance is called:**
   (a) Flute cast (b) Sole marks
   (c) Load cast (d) Dendrites

6. **Deepest part of the Cuddapah basin lies near the ____margin.**
   (a) Northern (b) Southern
   (c) Central (d) Western

7. **Main boundary fault separates:**
   (a) Lesser Himalayas from the outer Himalayas
   (b) Aravallis from Eastern Ghats
   (c) Gondwanas from Vindhyans
   (d) None of these

8. **During which era were the major physiographic features of India shaped?**
   (a) Proterozoic Era (b) Palaeozoic Era
   (c) Mesozoic Era (d) Cenozoic Era

9. Sitasoangi formation in Central India belongs to____ group of the rocks.

(a) Chilpi (b) Sakoli

(c) Saucer (d) Bijawar

10. Which of the following rocks is considered as a source rock for East Coast bauxite deposit in India?

(a) Kodurite (b) Gondite

(c) Khondalite (d) Charnockite

11. Which of the following rocks are present in the Marwar Supergroup of Rajasthan?

(a) Sandstone-Basalt-Mudstone

(b) Phyllite-Graywacke-Dolomitic-Limestone

(c) Sandstone-Banded gneiss-Marble

(b) Sandstone-Mudstone-Limestone

12. Lower Gondwana coal seams belong to which one of the following ages?

(a) Permo- Triassic (b) Late Carboniferous- Permian

(c) Triassic- Jurassic (d) Ordovician- Silurian

13. The Aravalli mountain are predominantly constituted by the rocks of a)

(a) Aravalli Supergroup (b) Bhilwara Supergroup

(c) Malani Igneous Suite (d) Delhi Supergroup

14. The oldest dated rock from so far of Rajasthan, is a –

(a) Dolerite (b) Conglomerate

(c) Gneiss (d) Quartzite

15. Which of the following stratigraphic units is not present in Rajasthan?

(a) Triassic (b) Jurassic

(c) Cretaceous (d) Eocene

16. Which of the following minerals can be used for determining the age of the Vindhyan Sandstone?

(a) Pitchbland (b) Monazite

(c) Glauconite (d) Ilmenite

17. Which of the following is the correct hierarchial order of lithostratigraphic units?

(a) Group-Formation-Member-Bed
(b) Formation-Group-Bed-Member
(c) Bed-Member-Formation-Group
(d) Group-Member-Formation-Bed

18. Which of the following is a plate boundary as well as Strike-slip fault?

(a) Himalayan Frontal Fault (b) Great Boundary Fault
(c) Great Glen Fault (d) San Andreas Fault

19. In which one of the following formation of Rajasthan good quantity of grow water is reported?

(a) Sanu Sandstone (b) Gotan Limestone
(c) Lathi Sandstone (d) Mandai Sandstone

20. The intertrappend beds of Deccan Traps are of ......origin.

(a) Volcanic (b) Marine
(c) Fluvio-lacustrine (d) Continental

21. Which of the following is a Devonian marker horizon in the Himalayas?

(a) Lipak formation (b) Muth Quartzite
(c) Po-formation (d) Syringothyris limestone

22. Match the coal field listed in List-I with their occurrences in states listed in List-II:

List-I	List-II
(1) Talchir	(i) Chhattisgrh
(2) Nayveli	(ii) Orissa
(3) Umariya	(iii) Tamil Nadu
(4) Jhilmili	(iv) Madhya Pradesh

Code:

(a) 1-(i), 2-(ii), 3-(iii), 4-(iv)
(b) 1-(ii), 2-(iii), 3-(iv), 4-(i)
(c) 1-(iii), 2-(iv), 3-(i), 4-(ii)
(d) 1-(iv), 2-(i), 3-(ii), 4-(iii)

23. Main boundary fault separates rocks belongs to which of the following?

(a) Upper Vindhyan and Aravalli
(b) Siwalik and Aravalli
(c) Siwalik and Tertiary
(d) Siwalik and Bundelkhand

24. Half-life of 87Rb is -

(a) 11.9 × 109 years (b) 50×109 years
(c) 14.01×109 years (d) 4.468×109 years

25. Which one of the following structures is of Bombay High?

(a) Doubly-plunging anticline
(b) Doubly-plunging syncline
(c) Non plunging anticline
(d) Non plunging syncline

26. On the index-to-sheets, the topo sheet 45 B/9 will be to the ________.

(a) North (b) East
(c) West (d) South

27. In India the superimposed drainage is well-developed in which region ?

(a) Eastern-Ghat (b) Marginal areas of Deccan Traps
(c) Indo-Gangetic Plain (d) In the Thar desert in NW India

28. The rock unit that were deposited during a specific interval of geologic time, are described as

(a) Interval zone (b) Facies zone
(c) Charonostratigraphic unit (d) Magnetic unit

29. The age of marine beds associated with Lower Gondwana rocks is :

(a) Early Triassic (b) Late Jurassic
(c) Early Cretaceous (d) Earliest Permian

30. Marine transgression of the coast of Kachchh began in:

(a) Cretaceous (b) Jurassic
(c) Paleocene (d) Oligocene

31. The age of Gondwana rocks in Kachchh is:

(a) Paleocene (b) Oligocene

(c) Early Cretaceous (d) Jurassic

32. Which of the following formation is distinguished by the presence of dinosaur egg shells of Maastrichtian age?

(a) Ariyalur Formation (b) Bagh Formation

(c) Bhuj Formation (d) Lameta Formation

33. The age of the Marine transgression of Cauvery basin, Tamil Nadu is :

(a) Cenomanian (b) Maastrichtian

(c) Damian (d) Sarmatian

34. Which of the following dating technique helps in establishing the chronology of glaciations?

(a) Potasium /Argon

(b) Rubidium/Stronatium

(c) Uranium/Lead

(d) Optically Simulated Luminescence

35. Lowering of ELA is indicative of:

(a) Recession of glacier (b) Rise of temperature

(c) Lowering of temperature (d) Negative Mass Balance

36. LGM refers to which time period?

(a) 110 Ka (b) 60 Ka

(c) 22 Ka (d) 200 Ka

37. Which of the following is not a proxy for inferring the Holocene palaeoclimate ?

(a) Lichenometry (b) Palynology

(c) Dendrochronology (d) Geochemistry of lake sediments

38. Some of the studies carried out show that the Glacial Maxima in Himalaya

(a) Has been synchronous throughout its Glacial Maxima in Himalaya

(b) Preceded by about 40,000 years as compared to global Maxima
(c) Lagged by about 40,000 years as compared to global Maxima
(d) Has been similar to global maxima

39. Deccan Basalt represents:

(a) A Plateau (b) A Mountain chain
(c) A Hillock (d) A Basin

40. The half-life in decay from 238U to 206Pd is:

(a) 4.5 billion years (b) 7.13 billion years
(c) 14.1 billion years (d) 5 3. billion years

41. Which rock is most reliable for dating, using the radioactivity of minerals?

(a) Coal (b) Metamorphic rocks
(c) Sedimentary rocks (d) Igneous rock

42. Which dating technique uses the annual growth of the tree rings?

(a) 14C (b) Dendrochronology
(c) Lichenometry (d) Fission track

43. The half-life of 14C is:

(a) 5730 years (b) 6000 years
(c) 7250 years (d) 8150 years

44. Half-life of 40K in Potassium Argon dating technique is:

(a) 1.0 billion years (b) 2.5 billion years
(c) 1.3 billion years (d) 4.2 billion years

45. The parent element for stable isotope 87Sr is:

(a) 238U (b) 235U
(c) 119U (d) 87 Rb

46. Which of the following is the most fundamental lithostratigraphic unit?

(a) Supergroup (b) Group
(c) Member (d) Formation

47. Age of Muth quartzite is

(a) Eocene (b) Cretaceous

(c) Triassic (d) Devonian

48. Which rock gives the age of its source material?

(a) Marble (b) Schist

(c) Sand stone (d) Slate

49. Which of the following is generally accepted as the time on the geologic time scale for the emergence of land plants is?

(a) Cretaceous (b) Triassic

(c) Permian (d) Silurian

50. The coal deposits of the east India belongs to which of the following time periods?

(a) Jurassic and Cambrian (b) Eocene

(c) Pliocene and Miocene (d) Permian

51. Of the following four geologic periods, which is the oldest?

(a) Devonian (b) Silurian

(c) Permian (d) Jurassic

52. Of the following four geologic periods, which is the most recent?

(a) Devonian (b) Silurian

(c) Permian (d) Jurassic

53. The Miocene, Eocene and Paleoceneepoches belong to what geologic period on the Phanerozoic Time Scale?

(a) Devonian (b) Permian

(c) Triassic (d) Tertiary

54. During which period, in the earth's history, did MOST dinosaurs become extinct?

(a) Carboniferous (b) Cretaceous

(c) Quaternary (d) Triassic

55. The extinction of the dinosaurs also marks the end of which geological era?

(a) Precambrian
(b) Cambrian
(c) Mesozoic
(d) Cenozoic

56. Which is the lowest formation in Jurassic of Kutch?

(a) Katrol
(b) Chari
(c) Umia
(d) Patcham

57. Which is the top most formation of Vindhayan Supergroup?

(a) Rewa formation
(b) Bhander formation
(c) Kaimur formation
(d) Semri formation

58. Which is the oldest rock group in Rajasthan?

(a) Banded Gneissic complex
(b) Aravalli Supergroup
(c) Delhi Supergroup
(d) Birmania formation

59. Term peninsula stands for:

(a) Tectonically stable
(b) Area covered mostly by metamorphic and igneous rocks
(c) Area surrounded by three sides by sea
(d) All of the above

60. Which is the most ancient mountain ranges in India?

(a) Himalaya
(b) Vindhyan
(c) Aravalli
(d) Cuddapah

61. Great boundary fault separates:

(a) Vindhyans from Pre-Vindhyans
(b) Kurnools from Cuddapahs
(c) Delhis from Aravalis
(d) None of the above

62. Main boundary fault separates :

(a) Upper Gondawana from Lower Gondwana
(b) Archeans from Delhi
(c) Siwalik from Lower Himalayas
(d) None of the above

63. Bap Boulder beds are found in:

(a) Himachal Pradesh (b) Uttar Pradesh

(c) Gujarat (d) Rajasthan

64. The general strike of Vindhayan mountain ranges is:

(a) N-S (b) E-W

(c) NE-SW (d) NNE-SSW

65. The general strike of Aravalli mountain ranges are:

(a) N-S (b) E-S

(c) NE-SW (d) NNW-SSE

66. Which one of the following minerals can be used for determining the age of the Vindhyan sandstone?

(a) Pitchblende (b) Monazite

(c) Glauconite (d) Ilmenite

67. The Great Boundary Fault (GBF) is present in which region –

(a) Sirohi - Barmer region (b) Kanoi – Siwana region

(c) Chittor – Kota region (d) Deedwana – Chouhtan region

68. Which Structure is considered as Impact Crater in Rajasthan ?

(a) Mer - Mundwara Ring Structure

(b) Sarnu - Dandali Igneous Forms

(c) Nagana Structure

(d) Ramgrah Structure (Baran district)

69. Consider the following statements:

(1) Stratigraphy deals with sequential arrangement of layered rocks according to the time of their formation.

(2) Correlation aims at matching the rock formations of distant areas deposited at the same stage of Earth's history.

(3) Correlation aims at matching the rock formations of distant areas deposited during different stages of Earth's history.

Which of these statements are correct?

(a) 1 only (b) 1 and 2

(c) 1 and 3 (d) 2 and 3

70. Consider the following statements:

(1) An Erathem is represented by a succession of rock formation deposited during an Era.

(2) A system does not constitute a part of an Erathem.

(3) A system is represented by a succession of rocks which were deposited during a period.

Which of these statements are correct?

(a) 1 only (b) 2 only

(c) 1 and 2 (d) 1 and 3

71. Which one of the following is the correct hierarchial order on lithostratigraphic unit?

(a) Group, Formation, Member and Bed

(b) Formation, Group, Bed and Member

(c) Bed, Member, Formation and Group

(d) Group, Member, Formation and Bed

72. Match List-I with List-II and select the correct answer using the codes given below the Lists:

	List-I		**List-II**
A.	Talchir Formation	1.	Mesoproterozoic
B.	Kajrahat Limestone Formation	2.	Lower Carboniferous
C.	Chinji Formation	3.	Middle Miocene
		4.	Archaean

Code:

	A	B	C
(a)	4	2	1
(b)	3	1	4
(c)	4	3	2
(d)	2	1	3

73. Which one of the following is the correct chronological order of different periods in the Paleozoic Era?

(a) Silurian, Cambrian, Ordovician, Permian, Carboniferous and Devonian

(b) Permian, Carboniferous, Devonian, Silurian, Ordovician and Cambrian

(c) Cambrian, Ordovician, Silurian, Devonian, Carboniferous and Permian

(d) Devonian, Carboniferous, Cambrian, Ordovician, Silurian and Permian

74. What is the age of Deccan Traps?

(a) Upper Cretaceous to Lower Eocene

(b) Lower Cretaceous to Upper Eocene

(c) Palaeocene to Lower Eocene

(d) Cambrian to Recent

75. Which one of the following represents the Physiographic divisions of India?

(a) Peninsula and Indo-Gangatic plains

(b) Peninsula, Indo-Gangatic plains and Extra-peninsula

(c) Extra-peninsula and Indo-Gangatic plains

(d) Peninsula and Extra-peninsula

76. On which basement the sedimen of Dharwar Supergroup were laid over?

(a) Marble and Conglomerate

(b) Phyllites and Slates

(c) Basalt and Conglomerate

(d) Sargur Schist and Peninsular Gnissic Complex

77. Which one of the following super-groups is known to be associated with diamonds?

(a) Vindhyan Supergroup

(b) Dharwar Supergroup

(c) Dongargarh Supergroup

(d) Gondwana Supergroup

78. Which one of the following rock types occur commonly in the Vindhyan Supergroup?

(a) Granites, Shales and Basalts

(b) Sandstones, Shales and Conglomerates

(c) Sandstones, Slates and Gneisses

(d) Conglomerates, Phyllites and Granites

79. Consider the following statements:

1. Damuda series of Gondwana contains a great store of mineral wealth in it's coal measures.
2. Most of the coal deposits of Jharia field belong to the Raniganj stage of Gondwana.
3. Besides coal, Manganese is the chief product of Damuda rocks.
4. The major coalfields of Gondwana rocks in India are in Bihar, Raniganj, Ballarpur and Singareni.

Which of these statements are correct?

(a) 1 and 2 (b) 2 and 3

(c) 3 and 4 (d) 1 and 4

80. Which one of the following is lithostratigraphic unit?

(a) Eratham (b) Acme zone

(c) Series (d) Formation

81. Which formation belongs to the lower Gondwana Sequences?

(a) Zawar (b) Poladpur

(c) Rohtas (d) Iron stone

82. Semri Group is related to the:

(a) Upper Vindhayan (b) Lower Gondwana

(c) Lower Vindhayan (d) Upper Gondwana

83. Which is not related to coalfield?

(a) Chandrapur (b) Sohagpur

(c) Ramgarh (d) Panna

84. Arranged in correct sequence from older to younger.

(a) Sargurschist complex, peninsula gneissic complex, Dharwar super group

(b) Patcham, Umia, Katrol, Charl

(c) Karharbari, Raniganj, Barakar, Kulti

(d) Papagani, Nallamalal, Cheyair, Kistna

85. Which one of the following separate the Eastern Dharwar to Western Dharwar?

(a) Peninsula gneissic complex (b) Bababudan

(c) Rani Bennur (d) Closepet granite

86. The maximum genera and Species in upper Gondwana sequence found in

(a) Mahadev (b) Jabalpur

(c) Rajmahal (d) Maleri

87. The general trend of Aravalli group is

(a) NW-SE (b) E-W

(c) ESE-WNW (d) NE-Sw

88. Dhokpathan formation is overlain by:

(a) Pinjor (b) Tatrot

(c) Nagri (d) Chinji

89. Match the following :-

	Column-A		**Column-B**
P.	Dhosa Oolite	1.	Gondwana Sequence
Q.	Mansar	2.	Dhawar supergroup
R.	Umaria Marine Bed	3.	Sausar group
S.	Hutti Goldfield	4.	Jurassic of Kutch

(a) P-3, Q-1, R-4, S-2 (b) P-2, Q-4, R-1, S-3

(c) P-1, Q-2, R-3, S-4 (d) P-4, Q-3, R-1, S-2

90. Dinosaur commonly found in which formation?

(a) Lameta Bed (b) Bagh Bed

(c) Deccan Trap (d) Rajmahal Trap

91. Rock salt is found is which place of India?

(a) Rajban (b) Rampur

(c) Mandi (d) Manali

92. The Fawn Limestone contains which one of the following?

(a) Colleniacolumnaris (b) Conophytoncylindricus

(c) Colleniasp (d) All of the above

93. The Jodhpur Sandstone is unconformably underlain by

(a) Colleniacolumaris (b) Conophytoncylindricus

(c) Colleniasp (d) All of the above

94. The Jodhpur sandstone is unconformably underlain by

(a) zone-stage-series-system-erathem

(b) series-zone-stage-system-erathem

(c) series-system-erathem-zone-stage

(d) zone-stage-system-series-erathem

95. The boundary between Mesozoic and Cenozoic is approximately at

(a) 55 Ma (b) 65 Ma

(c) 75 Ma (d) 125 Ma

96. Diamondiferous conglomerate occurs in which one of the following?

(a) Rewa group (b) Semri Group

(c) Bhander Group (d) Kaimur Group

97. The Salkhala Group is overlain by:

(a) Talchir formation (b) Dogra slates

(c) Mandhali formation (d) Gneissic rocks

98. In Kashmir, the Muth Quartzite is conformably overlain by:

(a) Fenestella Shales (b) Zewan formation

(c) Syringothyris limestone (d) None of the above

99. Which one is an Upper Gondwana Flora?

(a) Gangamopteris (b) Ptilophyllum

(c) Vertebraria (d) Glossopteris

100. Stegodon is a characteristic fossil of which formation?

(a) Kamalial formation (b) Pinjor formation

(c) Nagri formation (d) Dhokpathan formation

101. When did the first plants appear on land?

(a) Cambrian (b) Ordovician

(c) Silurian (d) Devonian

102. The Sargur Schist Complex is:

(a) older than Dharwan Group

(b) younger than Dharwar Group

(c) equivalent to closepet Granite

(d) younger than Papaghni group

103. The lava flow occupying large areas in western and central India is known as:

(a) Rajmahal trap (b) Rajahmundry trap

(c) Deccan trap (d) Sylhit trap

104. The fundamental unit of chronostratigraphic classification is:

(a) Erathem (b) System

(c) Stage (d) Series

105. Find odd one out:

(a) Period (b) Zone

(c) Age (d) Epoch

106. The Geological Time Scale (2004) includes:

(a) Two eons (b) Three eons

(c) Five eons (d) Seven eons

107. Which one is not a chronostratigraphic unit?

(a) System (b) Formation

(c) Stage (d) Series

108. The Chair Formation is best developed in

(a) Spiti basin (b) Kachchh basin

(c) Jaisalmer basin (d) Godavari basin

109. Isopach maps are used for analysis of

(a) Stratigraphic thickness (b) Depositional environments

(c) Sedimentary structures (d) Structural features

110. The Himalayan Neogene succession is represented by

(a) Dagshai Formation

(b) Siwalik Group and Karewa Formation

(c) Subathu Formation
(d) Kasuli Formation

111. The law of order of superposition of beds was proposed by:
(a) Steno (b) Smith
(c) Kiev (d) Philip

112. The Palaeozoic/Mesozoic boundary lies at:
(a) 270 ma (b) 251 ma
(c) 240 ma (d) 255 ma

113. Nallamalal Group is famous for:
(a) Iron (b) Pyrite
(c) Copper and lead (d) Mica

114. The strike of Aravalli is:
(a) NNW-SSE (b) NNE-SSW
(c) NE-SW (d) SW-NE

115. The presence of coal deposits impart great economic significance to
(a) Middle Gondwana sediments
(b) Lower Gondwana sediments
(c) Upper Gondwana sediments
(d) Late Triassic sediments

116. Find odd one out:
(a) Lathi formation (b) Chari formation
(c) Jaisalmer formation (d) Bhadasar formation

117. Palana lignite deposit is assigned to .... age.
(a) Late triassic (b) Tertiary
(c) Early jurassic (d) Cretaceous

118. The type locality for Triassic is in:
(a) England (b) Germany
(c) Canada (d) France

119. The community of interrelated organisms inhabiting in area is:

(a) Biotype (b) Thenatocoenose

(c) Biocenose (d) Thenotype

120. The lower Gondwana rocks are of ……….. age:

(a) Cambrian (b) Permian

(c) Jurassic (d) Triassic

121. The upper Gondwanas are characterized:

(a) Ptilopyllum flora (b) Glossopteris flora

(c) Gengamopten's flora (d) Dichroidium flora

122. The type of clay deposits associated with Gondwana rocks are:

(a) China clay (b) Fire clay

(c) Kaolin clay (d) None of these

123. In Odisha, coal deposits are associated with rocks of:

(a) Lower Gondwana (b) Upper Gondwana

(c) Tertiary (d) Gondwana

124. The Neotethys existed during:

(a) Paleozoic (b) Mesozoic

(c) Cenozoic (d) Neoproterozoic

125. Find odd one out:

(a) Lathi Formation (b) Chari Formation

(c) Katrol Formation (d) Umia Formation

126. The base of Palaeozoic Era is marked by first appearance of:

(a) Corals (b) Trilobites

(c) Pteropods (d) Cephalopods

127. The fundamental unit of chronostratigraphic classification is:

(a) Erathem (b) System

(c) Stage (d) Series

128. Mount Everest Limestone belongs to:

(a) Carboniferous (b) Ordovician

(c) Silurian (d) Devonian

129. The stratigraphic age of the Zewan beds is:

(a) Devonian (b) Permian

(c) Carboniferous (d) Cambrian

130. Isopach maps are used for analysis of:

(a) Stratigraphic thickness (b) Depositional environments

(c) Sedimentary structures (d) Structural features

131. The Himalayan Neogene succession is represented by:

(a) Dagshai Formation

(b) Subathu Formation

(c) Ranikot Formation

(d) Siwalik Group and Karewa Formation

132. The Chari Formation is best developed in:

(a) Spiti Basin (b) Jaisalmer Basin

(c) Kachchh Basin (d) Godavari Basin

133. Assign Kajrahat formation to the correct Group in the following:

(a) Kaimur (b) Bhander

(c) Semri (d) Rewa

134. Which one of the following is the fundamental unit of biostratigraphy?

(a) Barren zone (b) Abundance zone

(c) Bio zone (d) Overlap zone

135. Which one of the following groups contains Sirbu Shale Formation?

(a) Semri Group (b) Kaimur Group

(c) Rewa Group (d) Bhander Group

136. Gulcheru quartzite belong to which one of the following groups?

(a) Papaghani Group (b) Cheyair Group

(c) Nallamalai Group (d) Krishna Group

137. Which one of the following ages is assigned to Syringothyris Limestone?

(a) Ordovician (b) Lower Silurian

(c) Lower Carboniferous (d) Lower Permian

138. Which one of the following formations is correlated to Raniganj Formation in Son and Mahanadi valleys?

(a) Talchir Formation (b) Karharbari Formation

(c) Himgir Formation (d) Maleri Formation

ANSWER SHEET

1. (c)	2. (a)	3. (c)	4. (b)	5. (a)
6. (d)	7. (a)	8. (c)	9. (c)	10. (c)
11. (a)	12. (b)	13. (d)	14. (b)	15. (d)
16. (a)	17. (c)	18. (d)	19. (c)	20. (a)
21. (b)	22. (b)	23. (c)	24. (d)	25. (a)
26. (a)	27. (a)	28. (c)	29. (d)	30. (b)
31. (c)	32. (d)	33. (a)	34. (d)	35. (c)
36. (c)	37. (a)	38. (b)	39. (a)	40. (a)
41. (d)	42. (b)	43. (a)	44. (c)	45. (d)
46. (d)	47. (d)	48. (c)	49. (d)	50. (d)
51. (b)	52. (d)	53. (d)	54. (b)	55. (c)
56. (d)	57. (b)	58. (a)	59. (c)	60. (c)
61. (a)	62. (c)	63. (d)	64. (b)	65. (c)
66. (a)	67. (c)	68. (d)	69. (b)	70. (a)
71. (c)	72. (d)	73. (c)	74. (a)	75. (b)
76. (b)	77. (a)	78. (b)	79. (d)	80. (d)
81. (d)	82. (c)	83. (d)	84. (a)	85. (d)
86. (c)	87. (d)	88. (b)	89. (d)	90. (a)
91. (c)	92. (d)	93. (a)	94. (a)	95. (b)
96. (a)	97. (b)	98. (c)	99. (b)	100. (d)
101. (c)	102. (a)	103. (c)	104. (c)	105. (b)
106. (b)	107. (b)	108. (b)	109. (a)	110. (b)
111. (a)	112. (b)	113. (c)	114. (c)	115. (b)
116. (b)	117. (b)	118. (b)	119. (c)	120. (b)
121. (a)	122. (b)	123. (a)	124. (c)	125. (a)
126. (b)	127. (c)	128. (a)	129. (b)	130. (a)
131. (d)	132. (c)	133. (c)	134. (c)	135. (d)
136. (a)	137. (c)	138. (c)		

CHAPTER - 9

HYDROGEOLOGY

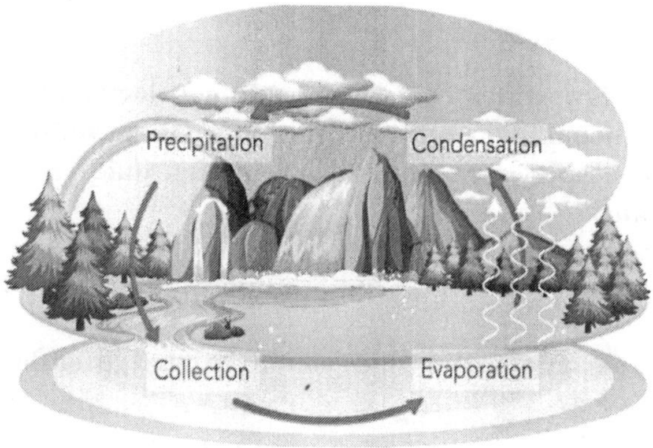

HYDROGEOLOGY

1. **Bivalent Fe ions precipitate at a pH value of:**
   (a) 3 (b) 4
   (c) 5 (d) 6

2. **Which basic law governs partition coefficient?**
   (a) Berthelot – Nernst equation
   (b) Clausius – Clapeyron equation
   (c) Henry's law
   (d) Clapeyron equation

3. **Which law states that the "settling velocity of a particle is proportional to the square of the partial diameter were density of particle. Acceleration due to gravity and fluid viscosity remains constant)"?**
   (a) Petti John's Law (b) Krumbein's Law
   (c) Stoke's Law (d) Snell's Law

4. **The property of rocks, which allows the passage for fluid movement is known as-**
   (a) Porosity (b) Absorption
   (c) Elasticity (d) Permeability

5. **The water present in the rocks from the time of their deposition in aqueous environment is names as-**
   (a) Juvenile water (b) Vadose water
   (c) Connate water (d) Cosmic water

6. **A rock body which may be porous enough to hold significant quantity of water but does not allow an easy and quick flow of water, is known as -**
   (a) Aquitard (b) Aquifuge
   (c) Aquiclude (d) Aquifer

7. **An impermeable rock through which there in no possibility of storage or movement of water is named as -**
   (a) Aquitard (b) Aquifuge
   (c) Aquifer (d) Aquiclude

8. The porosity of the unconsolidated clay in the range of -

(a) 0 to 5% (b) 5% to 10%

(c) 10% to 15% (d) 45% to 60%

9. It is desirable to use Groundwater than Surface water because:

(a) The temperature and chemical composition of groundwater is nearly constant

(b) Groundwater storage is not seriously affected by droughts

(c) It is sorted By Nature through many years of recharge

(d) All of these

10. Groundwater development is discouraged in some areas because:

(a) Underlain rocks have insufficient porosity and permeability to yield much water to wells

(b) TDS content in groundwater is greater than in surface water

(c) In regions of high precipitation cost of developing wells is higher than the cost of developing small streams

(d) All of these

11. In the vicinity of cities the TDS is higher in precipitation because-

(a) The dust and smoke produced by factories

(b) Dissolved gases

(c) There are no TDS in precipitation

(d) The dust and smoke produced by factories and dissolved urban gases

12. The amount of energy required to evaporate 1 cubic centimetre of water is

(a) 597 calories (b) 1000 calories

(c) 1597 calories (d) 100 calories

13. Vadose water is in the vadose zone that includes:

(a) Groundwater and capillary water

(b) Capillary water, gravitational water and soil water

(c) Capillary water, gravitational water and soil water

(d) Groundwater, gravitational water and soil water

14. Water table is:

(a) The surface in unconfined material along which the hydrostatic pressure is equal to the atmospheric pressure

(b) The surface separating the capillary fringe from the zone of saturation

(c) Water level surface in wells that tap an unconfined saturated material

(d) All of these

15. Aquitards:

(a) Are layers of clay and silt that related water flow underground and act as barriers for groundwater

(b) Direct the surface water that seeps down and replenishes aquifers

(c) Both statements are incorrect

(d) Both statements are correct

16. Water level fluctuations are due to:

(a) Changes in groundwater storage

(b) Defamation of aquifers

(c) Fluctuations of atmospheric pressure

(d) All of these

17. If there are 10-5 moles per litter of H^+ in water

(a) Then the pH of water is 5

(b) Then it is pure water

(c) Then water contains no dissociated OH^- ions

(d) Then the ph of water is 10

18. 63 PPM $Mg2^+$(At wt of Mg= 24.32, valency =2) is equal to:

(a) 5.19 epm (b) 6.7 epm

(c) 63 epm (d) 6.3 epm

19. Specific electric conductance of potable subsurface water ranges from:

(a) 32 to 2000 micromhos (b) 0.055 to 0.5 micromhos

(c) 45000 to 55000 micromhos (d) 5.0 to 30 micromhos

20. Water for domestic and industrial uses should have less than 1000 PPM TDS and for agriculture uses less than 3000 PPM TDS:

(a) True
(b) False
(c) About 300000 PPM
(d) 10 and 100 PPM TDS respectively

21. Water that has never been part of Hydrosphere is called:

(a) Juvenile water
(b) Ice water
(c) Metamorphic water
(d) Connate water

22. A mixture of miscible saltwater and freshwater is treated as homogeneous

(a) True
(b) False
(c) Only if density does not vary from place to place
(d) Is treated as heterogeneous

23. Porosity will be highest for cubicle array of

(a) Rhombs
(b) Spheres
(c) Cubes
(d) Rectangular sandstones pieces

24. Hydraulic conductivity depends on properties of the fluid as well as characteristics of the medium:

(a) Darcy failed to recognize the above fact
(b) Dupuit failed to recognize the above fact
(c) Both Darcy and Dupuit failed to recognize the above fact
(d) Darcy was aware of the above fact

25. Intrinsic permeability is characteristic of the

(a) Properties of soil and fluid
(b) Properties of soil (medium)
(c) Properties of fluid
(d) Properties of groundwater

26. Reynold's No. is given as; if V is specific discharge, D is characteristic length

(a) $N_R = VD/v$
(b) $N_R = Vv/D$
(c) $N_R = Dv/V$
(d) $N_R = V+D/v$

27. The amount of water used in storage released from a aquifer with unit cross-section under a unit decline of head is represented as:

(a) K
(b) S
(c) T
(d) A

28. For every meter (m) of fresh water above mean sea level, the thickness of the fresh water lens resting on the salt water is about:

(a) 35 m
(b) 36 m
(c) 40 m
(d) 37 m

29. The rocks generally have distinctive porosity and permeability and for that a geologist during exploration of subsurface water first seeks help by utilizing:

(a) Petrography
(b) Drilling
(c) Seismic methods of exploration
(d) Magnetic methods of exploration

30. Total water available for recharge, ease of recharge and quantity of groundwater discharge at the surface is determined by:

(a) Hydrologic methods of prospecting
(b) Dowsing
(c) Geological maps
(d) Atmospheric analysis

31. The average permeability of igneous and plutonic rocks with depth will generally

(a) Increase
(b) Decrease
(c) Not change
(d) Highly increase

32. Difference in yield in Wells of Igneous and metamorphic rock aquifers is because of:

(a) Differences of mineralogy
(b) Differences in degree of weathering
(c) Differences in degree of compaction
(d) None of these

33. Pathological organism will move efficiently in alluvial aquifers if:

(a) Soil is thin or absent over water bearing Aquifers

(b) Wells are unprotected

(c) There is storm runoff in wells

(d) All of these

34. In volcanic rock aquifers, following rocks will generally have high porosity but very low permeability:

(a) Basalts (b) Tuffs

(c) Dike rocks (d) Intrusive

35. In Cenozoic sediments the salinity of Connate water is:

(a) Less than that of sea water (b) More than that of sea water

(c) Equal to the sea water (d) Less than that of meteoric water

36. The reduction of artesian pressure in valley deposits will induce:

(a) Expansion of Aquifers

(b) Compaction of Aquifers

(c) Compaction of Aquifers and adjacent silts and clays

(d) Expansion of Aquifers and adjacent silts and clays

37. Chemical character of groundwater in large valleys of tectonic origin will generally be:

(a) Highly variable

(b) The same

(c) Always free of salinity induced from ancient sea water

(d) Always free of evaporitic solutions

38. The greatest danger of organic pollution is:

(a) In areas of extremely permeable aquifers

(b) In areas of improper waste-disposable facilities

(c) Not a problem in Wells within large valleys

(d) All of these

39. What type of water bearing material has the most uniform hydro geologic properties?

(a) Sandstone

(b) Conglomerate

(c) Evaporite

(d) Dune sand (Consolidate)

40. The average chemical quality of groundwater is poorer in desert regions because:

(a) Slow circulation of groundwater will slow the flushing of connate water

(b) Salt contained in rain will get concentrated at the surface

(c) Fine dust in soluble salts blown from playas

(d) All of these

41. Varying electrode spacing in resistivity survey helps in:

(a) Vertical investigation of an aquifer

(b) Noting changes in quality of groundwater

(c) Knowing impermeable formations

(d) All of these

42. The ground around electrodes during resistivity survey is moistened to:

(a) Establish proper contact with Earth

(b) To break the conductivity

(c) Decrease the salinity of the ground

(d) None of these

43. The Sub-Surface salt water boundaries become apparent on a resistivity depth curve due to:

(a) Increase in resistivity (b) Decrease in resistivity

(c) Straightening of the curve (d) Refraction of the curve

44. Seismic waves may be reflected or refracted:

(a) At interface where a velocity change occurs

(b) In homogenous formation

(c) In groundwater

(d) In ocean water

45. Resistivity Log is affected by

(a) Fluid within a well and ground water

(b) Character of the surrounding strata
(c) Well diameter
(d) All of these

46. In SP log, flow from the formation into the wells shows:

(a) Positive values
(b) Negative values
(c) No change in values
(d) Sharp change in negative value

47. To note water from different aquifers interested by a well, readings are taken of:

(a) SP log
(b) Resistivity log
(c) Temperature log
(d) Calliper log

48. Well diameters along a well can be measured with the help of:

(a) Calliper log
(b) Neutron log
(c) Temperature log
(d) Resistivity log

49. Tracer-techniques are adapted to note:

(a) Ground water velocity
(b) Ground water quality
(c) Ground water occurrence
(d) Ground water permeability

50. Which is used in Radioactive logs:

(a) Neutron rays
(b) Gamma particles
(c) Both Gamma Particles and Neutron rays
(d) All of these

51. The equation $h^2 = W/.0512K (R^2-r^2)$ shows depth to salt water at any location is function of:

(a) Rainfall recharge, size of the island and permeability
(b) Rainfall recharge and permeability
(c) Rainfall recharge and size of the island
(d) Size of the island and permeability

52. Modification in composition of sea water entering an aquifer can occur:

(a) By Base Exchange between water and minerals of aquifer
(b) Sulphate reduction, Solution and precipitation

(c) By Base Exchange between water and minerals of aquifer & Sulphate reduction, solution and precipitation

(d) None of these

53. One of the following does not have significant contribution in establishing Hydraulics of groundwater:

(a) Henry Darcy (b) N. Steno

(c) J.Dupuit (d) A.Theim

54. Initially dry soil that will absorb maximum moisture in contact with atmosphere is called as

(a) Wilting point (b) Capillary fringe

(c) Hydroscopic coefficient (d) Moisture point

55. Soil moisture is measured by neutron scattering based on the fact that fast neutrons are slowed by collision with:

(a) Hydrogen (b) Oxygen

(c) Chlorine (d) Water

56. As per Darcy's law flow rate through porous media is:

(a) Inversely proportional to head loss

(b) Not in accordance with hydraulic gradient

(c) Proportional to flow path length

(d) None of these

57. The water which results due to mixing of freshwater and sea water is:

(a) Saline water (b) Alkaline water

(c) Brackish water (d) Acidic water

58. Permeability K is expressed as Q/A (dh/dl), that show K has dimensions of:

(a) Velocity (b) Area

(c) Acceleration (d) Gravity

59. Equilibrium (Thiem) equation Q=2Kb h - hw/In (ro/rw)Enables determination of:

(a) Permeability from a pumped well

(b) Porosity from a pumped well

(c) Permeability from non-pumping well
(d) Transmissivity from non-pumped well

60. Theis equation ho-h = 114.6Q/T W(u),

Where W(u) is called as well function.

The argument (u) is given by

(a) $u=1.87r^2S/Tt$
(b) $u=2.87r^2S/Tt$
(c) $u=3.87r^2S/Tt$
(d) $u=4.87r^2S/Tt$

61. The resistivity of Fresh water s around '20 ohm m' it's conductivity therefore should be about:

(a) 500 Siemens/cm
(b) 500 microS/cm
(c) 500 microS/meter
(d) 500 microS/feet

62. The units of Yield, Transmissivity and drawdown are:

(a) m^3/s, m^2/s and m respectively
(b) m^2/s, m^2/s respectively
(c) m^3/s, m^2/s and m^2 respectively
(d) m/s, m^2/s and m respectively

63. Magnetic resonance sounding method for ground water exploration has many advantages such as: it is only geophysical method that is capable of estimating permeability but has disadvantage such as:

(a) Not suitable for volcanic rocks
(b) It is sensitive to power lines, pipes, fences
(c) The maximum depth which can be reached to detect an aquifer layer is 150 m
(d) All of these

64. Diseases which can spread through ground water are:

(a) Typhoid, cholera and Hypertension
(b) Typhoid, amoebic dysentery and diabetes
(c) Typhoid, amoebic dysentery, infectious hepatitis and malaria
(d) Typhoid, cholera, amoebic dysentery and infectious hepatitis

65. The acceptable concentration of fluoride in drinking water is about:

(a) 1.5 mg/L
(b) 1.5 g/L
(c) 1.5 μg/L
(d) 1.5 Kg/L

66. The acceptable level as defined by WHO for maximum concentrations of arsenic in safe drinking water is:

(a) 0.01 mg/L (b) 0.10 g/L
(c) 1.00 mg/L (d) 0.01 g/L

67. Arsenic contamination of the groundwater in Bangladesh around Ganga delta is a serious problem. It is one of the solutions is:

(a) To tap into deeper groundwater
(b) To tap into shallow groundwater
(c) To tap very shallow groundwater
(d) None of these

68. Chloride concentration in water can be reduced by using adsorbent such as:

(a) Activated charcoal and Alumina
(b) Bentonite and kaolinite
(c) Red mud and rice husk
(d) All of these

69. The groundwater problems being faced in Jodhpur are:

(a) Declining water level, over exploration of groundwater
(b) Fluoride and nitrate pollution
(c) Rising water level, leakage through sewerage system
(d) All of these

70. Rooftop rainwater harvesting has been made mandatory in state old buildings of plot size more than

(a) 300 sq.m (b) 500 sq.m
(c) 600 sq.m (d) 200 sqm

71. Tick mark the correct statement

(a) Recharge to aquifer during normal rainfall in western Rajasthan is inadequate
(b) Major portion of precipitation is in western Rajasthan gets lost as run off
(c) Identification of potential aquifers is still needed in western Rajasthan
(d) All of these

72. The advantage of watershed development are:

(a) Arrests soil erosion, improves soil moisture and recharges groundwater

(b) Reclaims eroded land and reduces flood

(c) Arrests soil erosion, improves soil moisture and recharges groundwater and reclaims eroded land reduces flood, both are correct

(d) Only arrests soil erosion, improves soil moisture and recharges groundwater is correct

73. How much of Earth's water is stored in underground aquifers?

(a) Less than 1% (b) 2%

(c) 9.5% (d) 5%

74. An area of land that contains a common set of streams and rivers that all drain into a single larger body is called watershed for example:

(a) Mississippi river (b) Jaisalmer lake

(c) Stream and ocean (d) All of these

75. Rainfall is the main source of groundwater recharge, the average annual rainfall in Western Rajasthan is about:

(a) 10 cm (b) 30 cm

(c) 100 cm (d) 1000 mm

76. A confined aquifer has transmissivity of 40 m^2/day:

The slope of the piezometric surface is 0.25 m/km.

How much water per day flow through an aquifer

Per kilometer width of the aquifer? (hint Q=WTi)

(a) 100 m^3/day (b) 10 m^3/day

(c) 1 m^3/day (d) 1000 m^3/day

77. Radioactive waste disposal can cause groundwater pollution. Possible sites for radioactive waste disposal are:

(a) Salt domes and brine aquifers

(b) Subterranean cavity created by underground nuclear explosion

(c) Deep mines

(d) All of these

78. Groundwater contamination due to coal mining is caused by:

(a) Oxidation of pyrite

(b) Formation of sulfuric acid and iron hydroxide

(c) Decrease in pH due to sulphate concentration

(d) All of these

79. Boris and Jhalra's like traditional water bodies of Western Rajasthan are:

(a) Surface and groundwater bodies

(b) Only surface water bodies

(c) Only ground water bodies

(d) None of these

80. Remote Sensing for groundwater involves sensing of:

(a) Energy source and radiation (b) Interaction with targets

(c) Atmospheric interaction (d) All of these

81. Exploration by resistivity method helps in assessing the quality of groundwater because

(a) Resistivity increases with more dissolved solid in water

(b) Distilled water has very high conductivity

(c) Resistivity decreases with increasing amount of ions

(d) Water is not a very good electrolyte

82. Free air correction is applied in gravity method of exploration because:

(a) Major value of gravity decreases with the increase in the elevation

(b) Measure value of gravity increases with the increase in the elevation

(c) The equation g0=GM/R2 is not correct

(d) Isostatic correction is not required in geodetic survey

83. The magnetic minerals such as magnetite ilmenite pyrrhotite etc locally influences the Earth's field depending upon a factor called:

(a) Dielectric constant (b) Conductivity

(c) Permeability (d) Resistivity

84. The variation in the velocity of Elastic seismic waves is dependent upon the Rock type and its degree of consolidation therefore:

(a) Their velocity will not very for the same Rock type such as shale

(b) Their velocity will vary for the same Rock types such as shale

(c) Their velocity will not vary for the same Rock types such as granite

(d) Their velocity will not verify the same Rock types such as sandstone

85. To determine accurate gift to determine accurate Depth of the beds specially to locate structural traps for groundwater, the exploration method employed is:

(a) Electromagnetic method (b) Gravity method

(c) Seismic reflection method (d) Magnetic method

86. Mark the true statement:

(a) Compressed air is used in pneumatic drills

(b) Percussion drill cannot be used for making blast hole

(c) Hammer drills cannot be used as hand held drills

(d) Diamond drills are advisable in the soft Rock formations

87. Ground penetrating radar (GPR) and seismic reflection techniques are analogues because

(a) Both techniques record the time required for a wave to travel to the interface between two formations and then return to the surface

(b) Both techniques record the velocity required for a wave to travel to the interface between two formations and then return to the surface

(c) Both techniques do not record the time required for a wave to travel to the interface between two formations and then return to the surface

(d) Both techniques do not record the velocity required for a wave to travel to the interface between two formations and their return to the surface

88. If the pumping rate is 1000 m^3/day and the drawdown is 30 m

(a) The specific capacity of the well is 33.3 m^3/day/m

(b) The transmissivity of the well is 33.3 m^3/day/m

(c) The storage coefficient of the well is 33.3 m^3/day/m

(d) All of these

89. For industrial Municipal or irrigation purpose the transmissivity of aquifer should generally be more than:

(a) 125 m^2 /day/m
(b) 125 m^3 /day/m
(c) 125 m^3 /day
(d) 225 m^2 /day

90. The flow of groundwater to the well forms a cone of depression. The volume of cone is:

(a) 1/3 $4\pi r^3 h$
(b) 1/3 πr^3
(c) 1/3 $\pi r^3 h$
(d) 4/3 πr^3

91. Groundwater flow velocity can be measured with the help of:

(a) Cannot be measured
(b) Tracers
(c) Dowsing
(d) It remains constant

92. The flow net is:

(a) A set of interesting flow lines and equipotential line
(b) A set of meandering river lines
(c) A set of wind blowing directions
(d) A set of groundwater velocity lines

93. Water having more than 10,0000 mg/ITDS is called:

(a) Saline water
(b) Brackish water
(c) Brine water
(d) Freshwater

94. Radionuclide contamination in groundwater is measured in:

(a) PPM
(b) PPB
(c) PCi/l
(d) gm/l

95. The energy responsible for generating tornadoes and hurricane comes from water because it has

(a) High heat capacity
(b) Highest heat of vaporization
(c) Highest solubility
(d) Quick dissociation

96. The source of oregano-mercury compounds as pollutants in groundwater are:

(a) Gasoline
(b) Fungicides
(c) Fertilizers
(d) Septic tanks

97. The source of late in groundwater as pollutant is

(a) Gasoline (b) Fungicides
(c) Fertilizers (d) Septic tanks

98. Ozone in the atmosphere is destroyed by:

(a) Photo dissociation by UV rays
(b) Reaction with NO
(c) Reaction with chlorofluorocarbons
(d) All of these

99. Relative humidity is:

(a) Amount of water vapor per m^3 of air
(b) % of maximum amount of water vapor that air can hold
(c) Amount of water vapor per m^3 of air and % of maximum amount of water vapor that air can hold, both are correct
(d) None of these

100. The Ocean have an average salinity of 34.77% while freshwater has:

(a) 0% salinity (b) Salinity of 0.2 to 4%
(c) 0.2% salinity (d) 4% salinity

101. Monsoon is the wind system that:

(a) Never fluctuates
(b) Maintains its course in the opposite season
(c) Reverses its course in the opposite season
(d) Has nothing to do with the rotation of the earth

102. Which feature is a sign of Karst topography?

(a) Sinkholes (b) Artesian Wells
(c) Moraines (d) Anticline

103. During ice ages the sea level was:

(a) 5 km lower than today (b) Same as today
(c) Higher than today (d) 120 m lower than today

104. Water is critical in maintaining our uniform body temperature because it has

(a) High heat capacity (b) Highest heat of vaporization
(c) Highest solubility (d) Quick dissociation

105. Water in the atmosphere:

(a) Radiates back heat received from the Sun

(b) Cannot radiate back heat received from the Sun

(c) Has no effect on heat received from the Sun

(d) Makes the sun look brighter

106. Atmospheric Windows useful in Remote Sensing are created because:

(a) Ozone absorbs UV radiation from the Sun

(b) Carbon dioxide absorbs radiation in the infrared portion of the spectrum

(c) Water vapor absorbs portions of the infrared and microwave radiation

(d) All of these

107. Horizontal permeable conduit for collecting ground water by gravity flow is known as:

(a) Tube well (b) Infiltration gallery

(c) Gravity gallery (d) Dug well

108. High amount of sodium in water (sodium hazard) causes:

(a) Reduction in soil permeability

(b) Hardening of soil

(c) Reduction in soil permeability and hardening of soil, both are correct

(d) Only (hardening of soil) is correct

109. In electrical resistivity method user for ground water investigation, a current is generated in the field of frequency -

(a) Less than 1 cycle per second

(b) Less than 5 cycle per second

(c) More than 5 cycle per second

(d) More than 10 cycle per second

110. In which one of the following aquifers the storage coefficient is equal to specific yield?

(a) Confined aquifer

(b) Unconfined aquifer

(c) Confined and unconfined aquifer both
(d) Artesian aquifer

111. Which one of the following does not form a good aquifer?

(a) Jointed basalt (b) Massive granite
(c) Tills (d) Sandstone

112. A rock that is porous but not permeable:

(a) Marble (b) Sandstone
(c) Silt stone (d) Granite

113. A rock which is porous and impermeable is called:

(a) Aquifer (b) Aquitard
(c) Aquifuse (d) Aquiclude

114. The water that is entrapped in sedimentary rocks during their formation is called:

(a) Connate water (b) Juvenile water
(c) Vadose water (d) Meteoric water

115. Out of the following rocks which one is a good aquifer?

(a) Sand (b) Clay
(c) Silt (d) Shale

116. The ground water zone where pore spaces are completely saturated with water is known as:

(a) Zone of aeration (b) Zone of saturation
(c) Zone of subsurface (d) Zone of internal water

117. The surface separating the zone of saturation from the zone of accretion is called:

(a) Piezometric surface (b) Water table
(c) Mean sea surface (d) None of these

118. The water movement property of a rock is given by:

(a) Permeability (b) Porosity
(c) Specific yield (d) Specific retention

119. The linear law of movement of water was given by:

(a) White
(b) Linnaeus
(c) Zone of internal water
(d) None of these

120. The perched water lies in the:

(a) Zone of aeration
(b) Zone of saturation
(c) Zone of internal water
(d) None of these

121. Newly deposited clays have porosity:

(a) Up to 5%
(b) Up to 100%
(c) Up to 70%
(d) Up to 30%

122. The percentage of interconnected pore volume to the bulk rock volume is known as:

(a) Absolute porosity
(b) Effective porosity
(c) Fracture porosity
(d) Secondary porosity

123. Which of the following is true in a reservoir under hydrodynamic equilibrium?

(a) Water column is above oil column
(b) Gas column is above oil column
(c) Oil column is above gas column
(d) None of the above

124. API gravity is a measure of:

(a) Viscosity of the fluid
(b) Density of the fluid
(c) Porosity of the rock
(d) Permeability of the rock

125. Hydraulic fracturing of a well will generally:

(a) Decrease the average permeability near the wellbore
(b) Damage the formation
(c) Improve the production of the well
(d) Produce more water

126. In an under saturated reservoir, the reservoir pressure is:

(a) Equal to bubble point pressure
(b) Less than the bubble point pressure
(c) More than the bubble point pressure
(d) None of these

127. Which type of the land cover will provide the highest degree of recharge in a watershed?

(a) Pasture of grazing land
(b) Forest with thin carpet of litter
(c) Forest with thick carpet of litter
(d) Cropland

128. The ability of rocks or sediments to contain water is determined by its:

(a) Permeability
(b) Porosity
(c) Viscosity
(d) Transmissivity

129. A geological formation that holds enough water and transmits it rapidly enough to be useful as a source of ground water is known as:

(a) Aquifer
(b) Aquitard
(c) Aquifuse
(d) Aquiclude

130. Shallow foundations are sensitive to the vertical displacement component, especially if the structure is:

(a) Heavy and water table is shallow
(b) Light and water table is shallow
(c) Heavy and water table is deep
(d) Light and water table is deep

131. Which one of the following satellites can be used for monitoring changes in ground water storage?

(a) LANDSATTM
(b) IRS LISSIII
(c) Quick Bird
(d) GRACE

132. How much water the ice caps and glaciers contribute in terms of percentage at or near the Earth's surface?

(a) 2.05 %
(b) 3.05 %
(c) 1.78 %
(d) 1.23 %

133. What type of well can be constructed for small domestic water supplies in clayeysandy soil region?

(a) Deep and small diameter well
(b) Shallow and large diameter well

(c) Shallow and small diameter well
(d) Deep and medium diameter well

134. What is the nature of water table contour map in the region of ground water recharge?

(a) Convex water table contours
(b) Concave water table contours
(c) Closely packed water table contours
(d) Sparsely packed water table contours

135. Clay lenses, occurring in alluvial formations which can produce water table conditions in the unsaturated zone of the main unconfined aquifer are known as:

(a) Perched aquifer
(b) Leaky aquifer
(c) Unconfined aquifer
(d) Composite aquifer

136. What is the name of the wide belt (5–15 Km) along the northern margin of Indo-Gangetic- Brahmaputra plain, where major artesian aquifers give rise to free flow in wells?

(a) Fore deep basin
(b) Tarai belt
(c) Bhabar belt
(d) Siwalik belt

137. Which mineral provides highest swelling property to the soil?

(a) Kaolinite
(b) Montmorillonite
(c) Ilmenite
(d) Quartz

138. What happens to soil quality if there is high sodium ion concentration in irrigation water?

(a) Increase in permeability of soil
(b) Decrease in permeability of soil
(c) Compaction of the soil
(d) Change in grain size of soil

139. If the concentration of Ca^{2+}in water is 130 mg/L, what will be the concentration of Ca^{2+}in meq/L?

[Atomic weight of Ca = 40 .08]

(a) 6.48 meq/L
(b) 6.80 meq/L
(c) 13.0 0 meq/L
(d) 3.24 meq/L

140. Which one of the following is the correct order of cation exchangeability for common ion in groundwater?

(a) $K^+ > Mg^{2+} > Ca^{2+} > Na^+$ (b) $Ca^{2+} > Mg^{2+} > K^+ > Na^+$

(c) $Na^+ > K^+ > Ca^{2+} > Mg^{2+}$ (d) $Ca^{2+} > K^+ > Mg^{2+} > Na^+$

141. According to Indian Standard, what is the maximum permissible limit of F^-(Fluoride) in groundwater for drinking purpose?

(a) 1.0 mg/L (b) 1.3 mg/L

(c) 1.5 mg/L (d) 2.5 mg/L

142. Which one of the following will exhibit the maximum diurnal fluctuation in groundwater table due to evapotranspiration?

(a) Bare soil (b) Shallow rooted vegetation

(c) Deep rooted vegetation (d) Exposed bed rock within soil

143. Which one of the following is the characteristic of piezometric surface of flowing well?

(a) Piezometric surface lies above the ground surface

(b) Piezometric surface lies below the ground surface

(c) Piezometric surface lies below the confined aquifer

(d) Piezometric surface lies below the mean sea level

144. In which type of unconsolidated sediments, the capillary rise of ground water is the highest?

(a) Fine gravel (b) Medium sand

(c) Coarse sand (d) Silt

145. A soil sample has d60 and d10 values of 1.9 mm and 0.2 mm respectively. The uniformity coefficient of the soil will be:

(a) 9.5 (b) 0.1

(c) 0.38 (d) 0.15

146. Which one of the following particles contributes to the greatest amount of surface area in unconsolidated formations?

(a) Gravel (b) Coarse sand

(c) Silt (d) Clay

147. In Intermediate Vadose zone, water moves:

(a) Under the influence of gravity
(b) Under the influence of groundwater head difference
(c) Under the influence of atmospheric pressure
(d) Under the influence of osmotic pressure

148. Which one of the following formations shows the maximum anisotropism and heterogeneity in terms of its hydrogeological properties?

(a) Crystalline rock formation
(b) Alluvial formation
(c) Lacustrine formation
(d) Eolian formation

149. A saturated but poorly permeable stratum that impedes groundwater movement and does not yield water freely to the wells is known as:

(a) Aquifuge (b) Aquiclude
(c) Aquitard (d) Aquifer

150. What type of information can be provided by Radar imagery of a region?

(a) Colour of the soil
(b) Presence of moisture at shallow depth
(c) Void ratio of the soil
(d) Texture of the soil

151. The recharge of the ground water can be improved by stream channel through:

(a) Proper lining at the bottom of channel
(b) Narrowing of the channel
(c) Construction of low check dam and dikes across the stream at wide stretch of a channel
(d) Channel bank development

152. The combination of specific yield and specific retention is known as:

(a) Transmissivity (b) Storativity
(c) Porosity (d) Hydraulic conductivity

153. The topographic area that collects and discharges surface stream flow through one outlet or mouth is generally referred to as:

1. Catchments
2. Drainage basins
3. Watersheds

Select the correct answer using the code given below:

(a) 1 and 2 only (b) 1 and 3 only
(c) 2 and 3 only (d) 1, 2 and 3

154. Which of the following groundwater management models is used for water allocation plans involving economic management objectives subject to institutional policies as constraints in addition to hydraulic management constraints?

(a) Embedding approach
(b) Groundwater policy evaluation
(c) Groundwater policy evaluation and allocation models
(d) Optimal allocation approach

155. Which of the following statements about juvenile water is/are correct?

1. It is surface water that was trapped by soil
2. It is subsurface water that was derived from the interior of the Earth
3. It is surface water that was trapped by rock formation
4. It is subsurface water that was trapped by soil and rock formation

Select the correct answer using the code given below:

(a) 1 only (b) 2 only
(c) 1 and 2 (d) 3 and 4

156. Which one of the following methods is NOT used for measurement of evapotranspiration?

(a) Priestley- Taylor method (b) Blaney- Criddle method
(c) Theissen method (d) Adjusted Pan method

157. The incidence of infantile "Methemoglobinemia" is caused due to intake of water having high concentration of:

(a) Boron (b) Fluoride
(c) Nitrate (d) Manganese

158. The ratio of transmissivity to the coefficient of storage (T/S) of an aquifer is termed as:

(a) Leakage factor (b) Hydraulic resistance
(c) Hydraulic diffusivity (d) Boulton delay index

159. Geysers and fumaroles are manifestations of hydrothermal phenomena and are nearly always found in:

(a) Regions of seismic activity
(b) Regions of volcanicity
(c) Regions of landslides
(d) Regions of plate collisions

160. Consider the following statements regarding hydrologic properties of sedimentary rocks:

1. It depends largely on the size, shape and arrangement of the grains
2. It depends largely on the degree to which the rock particles are sorted
3. It depends largely on hydraulic resistance, which of the statements given above is/are correct?

(a) 1 only (b) 2 only
(c) 1, 2 and 3 (d) 1 and 2 only

161. In the electrical resistivity method, if a material of resistance has a cross-sectional area and length, then its resistivity ( ) can be expressed as:

(a) = _____ (b) = ______
(c) = _____ (d) = _______

162. The method of prospecting which is based on anomaly in chemical constituents of natural water is termed as:

(a) Hydrological prospecting (b) Mineralogical prospecting
(c) Chemical prospecting (d) Hydrogeochemical prospecting

163. In which one of the following geophysical methods, the potential electrodes are located at one-third points between the current electrodes?

(a) Schlumberger arrangement (b) Wenner arrangement
(c) Gravity method (d) Magnetic method

164. Consider the following statements about well logging:

1. Clayey formations display lower resistivities than permeable alluvial aquifers
2. Clay minerals do not conduct electrical current through their matrix
3. In porous formations, water content and quality factors control their resistivity value, Which of the statements given above is/ are correct?

(a) 1 and 3 (b) 1 only

(c) 1 and 2 (d) 3 only

ANSWER SHEET

1. (c)	2. (a)	3. (c)	4. (d)	5. (c)
6. (c)	7. (b)	8. (d)	9. (d)	10. (d)
11. (d)	12. (a)	13. (c)	14. (d)	15. (d)
16. (d)	17. (a)	18. (a)	19. (a)	20. (a)
21. (a)	22. (a)	23. (b)	24. (c)	25. (b)
26. (a)	27. (b)	28. (c)	29. (a)	30. (a)
31. (a)	32. (b)	33. (d)	34. (b)	35. (a)
36. (c)	37. (a)	38. (d)	39. (d)	40. (d)
41. (d)	42. (a)	43. (b)	44. (a)	45. (d)
46. (a)	47. (c)	48. (a)	49. (a)	50. (d)
51. (a)	52. (c)	53. (b)	54. (c)	55. (a)
56. (d)	57. (c)	58. (a)	59. (a)	60. (a)
61. (b)	62. (a)	63. (d)	64. (d)	65. (a)
66. (a)	67. (a)	68. (d)	69. (c)	70. (a)
71. (d)	72. (c)	73. (a)	74. (d)	75. (b)
76. (b)	77. (d)	78. (d)	79. (a)	80. (d)
81. (c)	82. (a)	83. (c)	84. (b)	85. (c)
86. (a)	87. (a)	88. (a)	89. (a)	90. (c)
91. (b)	92. (a)	93. (c)	94. (c)	95. (b)
96. (b)	97. (a)	98. (d)	99. (a)	100. (c)
101. (c)	102. (a)	103. (d)	104. (a)	105. (a)
106. (d)	107. (b)	108. (c)	109. (d)	110. (b)
111. (b)	112. (b)	113. (d)	114. (a)	115. (a)
116. (b)	117. (b)	118. (a)	119. (c)	120. (a)
121. (c)	122. (d)	123. (b)	124. (d)	125. (c)
126. (c)	127. (b)	128. (b)	129. (a)	130. (c)
131. (d)	132. (a)	133. (c)	134. (a)	135. (a)
136. (c)	137. (b)	138. (b)	139. (d)	140. (d)
141. (c)	142. (b)	143. (a)	144. (d)	145. (a)
146. (d)	147. (a)	148. (a)	149. (c)	150. (b)
151. (c)	152. (c)	153. (d)	154. (c)	155. (b)
156. (c)	157. (c)	158. (a)	159. (b)	160. (c)
161. (b)	162. (d)	163. (a)	164. (a)	

CHAPTER - 10

REMOTE SENSING

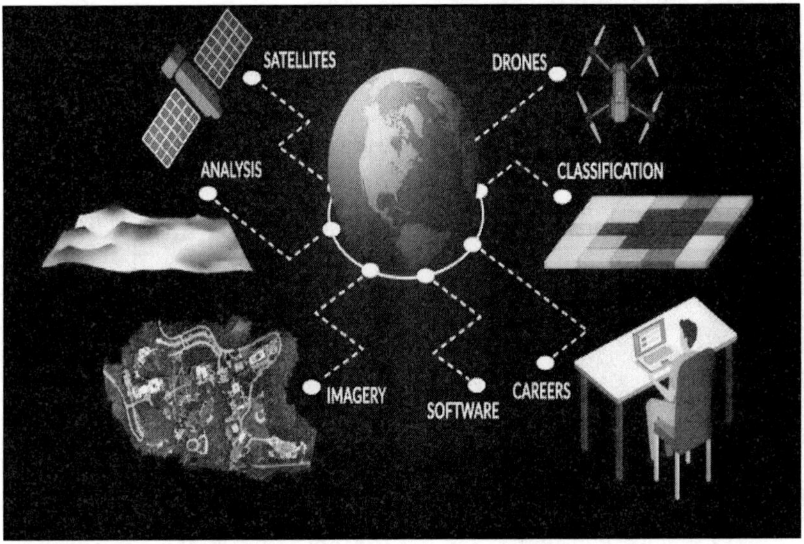

Remote Sensing

1. **What is the colour of a healthy vegetation in a false colour Infra-red positive print?**

   (a) Light Green (b) Dark Green
   (c) Blue (d) Red

2. **Assume that the relief of displacement for a tower 'A' is 3 mm and the radial distance from the centre of the photo to the top of the tower is 50 mm. If the flying altitude is 1350 m. above the base of the tower, find the height of the tower.**

   (a) 61 m (b) 56 m
   (c) 76 m (d) 81 m

3. **Which of the following is a geostationary satellite?**

   (a) Landset (b) INSAT
   (c) IRS (d) SPOT

4. **Geostationary satellites are launched in an orbit at a distance of:**

   (a) 32000 Km (b) 34000 Km
   (c) 36000 Km (d) 38000 Km

5. **"Pixel" stands for:**

   (a) Picture element (b) Photo element
   (c) Picture elevation (d) Photo elevation

6. **The source of Near Infrared part of "EMR" is:**

   (a) Sun (b) Objects on Earth
   (c) Satellites (d) Moon

7. **Atmospheric window for Remote Sensing is Part of EMR:**

   (a) Which passes through the atmosphere?
   (b) Which is absorbed by the atmosphere?
   (c) Which is absorbed by the water vapour?
   (d) None of these

8. Data acquisition systems which is acquired on different dates is called

(a) Multi-stage sensing (b) Multi-band sensing
(c) Multi-spectral sensing (d) Multi-temporal sensing

9. Spectral signature of an object is the value of EMR

(a) Absorbed by the object in different well defined bands
(b) Reflected by the object in different well defined bands
(c) Refracted by the object
(d) None of these

10. What is the average height of GPS satellite from earth surface –

(a) 36,100 km (b) 20,200 km
(c) 40,000 km (d) 45,000 km

11. The abbreviation DEM in the remote sensing stand for –

(a) Digital Electro Magnetism
(b) Digital Elevation Map
(c) Digitally Enhanced Measurement
(d) Digital Exponential Magnitude

12. Name the first Indian satellite:

(a) Rohini (b) Bhaskar
(c) INSAT-1 (d) Aryabhatta

13. The arrangement of terrain features, which provide shape, size, and textural attributes is called as -

(a) Temporal Variation (b) Spectral Variation
(c) Apatial Variation (d) Altitudinal Variation

14. The capability of the sensor to discriminate the smallest object on the ground of different sizes, usually specified in linear dimension, is known as:

(a) Spatial resolution (b) Spectral resolution
(c) Radiometric resolution (d) Temporal resolution

15. Which one of the following is NOT a spatial data?

(a) Location of the city (b) Location of the river
(c) Temperature of the city (d) Location of the hill

16. Which one of the following is NOT related to topology?

(a) Adjacency (b) Containment

(c) Connectivity (d) Attribute data

17. Which one of the following non-spatial scales would you select for recording of snow-depth in GIS database?

(a) Nominal scale (b) Ratio scale

(c) Ordinal scale (d) Interval scale

18. What is the spatial resolution of LISS III camera of IRS 1C?

(a) 72.0 m (b) 36.0 m

(c) 23.5 m (d) 6.0 m

19. Consider the following statements regarding conservative ions:

1. Conservative ions are those that will undergo association and dissociation reactions over the normal range of pH of natural water
2. Conservative ions are those whose concentrations are not affected by changes in pH, temperature or pressure, assuming no precipitation or dissolution
3. Na^+, K^+, Ca^{2+}, Mg^{2+}, Cl–, SO_4^+ and 3NO –are not the conservative ions of natural water Which of the statements given above is/are correct?

(a) 1 only (b) 2 and 3

(c) 2 only (d) 1 and 3

20. The over shaped structural feature in aerial Photograph indicates:

(a) Unconformity (b) Joint

(c) Doubly plunging fold (d) Dome

21. The components of GIS is:

(a) Hardware (b) Software

(c) Data (d) All of these

22. The element of aerial photo interpretation is:

(a) Tone (b) Texture

(c) Shape (d) All of these

23. Deforestation is occurring in many countries and it is therefore difficult to detect the ongoing damage. Which of the following

types of remote sensing would be best suited for locating deforestation?

(a) Thermal infrared (b) Microwave
(c) Radar (d) Colour infrared

24. Using the latitude and longitude system (degree, minutes, second), 20 minutes is equal to what?

(a) 0.20° (b) 1/3 hour
(c) 200 second (d) 1/3°

25. An automated system for the capture, storage, retrieval, analysis and display of spatial data is known as-

(a) GPS (b) LANDSAT
(c) GIS (d) None of the above

26. Precise measurement of Earth's features can be obtained from:

(a) High-oblique photographs
(b) Vertical aerial photographs
(c) Low-oblique photographs
(d) All the above types of photographs

27. Which of the following remote sensing technologies uses sound?

(a) Radar (b) Colour infrared
(c) Thermal infrared (d) Sonar

28. Who is said to be the "Father of the Indian Space Program"?

(a) Dr. A. P. J. Abdul Kalam (b) Dr. Vikram Sarabhai
(c) Dr. K. Kasturirangan (d) Prof. Satish Dhawan

29. The largest scale of the following is:

(a) 1:24000 (b) 1:62500
(c) 1:100000 (d) 1:500000

30. Which one is not an open source GIS technique:

(a) Quantum GIS (b) Arc GIS
(c) SAGA GIS (d) GRASS GIS

31. Which of the following is a reverse magnetic polarity event:

(a) Reunion (b) Olduvai
(c) Jaramillo (d) None of the above

32. Which of the following is a radio-isotope:

(a) 12C (b) 13C

(c) 14C (d) 18O

33. Carbon-14 has a half-life of:

(a) 6,000 years (b) 4,730 years

(c) 5,730 years (d) 10,000 years

ANSWER SHEET

1. (d)	2. (d)	3. (b)	4. (c)	5. (a)
6. (a)	7. (a)	8. (d)	9. (b)	10. (b)
11. (b)	12. (d)	13. (b)	14. (a)	15. (c)
16. (d)	17. (d)	18. (c)	19. (c)	20. (c)
21. (d)	22. (d)	23. (d)	24. (d)	25. (c)
26. (b)	27. (d)	28. (b)	29. (a)	30. (b)
31. (d)	32. (c)	33. (c)		

CHAPTER - 11

MINING AND ENGINEERING

MINING AND ENGINEERING

1. **A Dam is an example of which type of system:**
   (a) Natural (b) Modified
   (c) Controlled (d) Uncontrolled

2. **'Smith's test' is performed to determine which of the following properties of building stone?**
   (a) Durability (b) Frost resistance
   (c) Density (d) Porosity

3. **In case of arch dam the whole or greater part of the load is transferred to the:**
   (a) Foundation (b) Abutments
   (c) Reservoir water column (d) At the toe of dam

4. **Which type of clay in used as drilling mud?**
   (a) China Clay (b) Fuller's Earth
   (c) Bentonite Clay (d) Fire Clay

5. **Churn drill is a type of -**
   (a) Percussion drilling (b) Rotary drilling
   (c) Jet drilling (d) None of the above

6. **The principal of churn drilling is -**
   (a) Rotary action (b) Rotary action with water jet
   (c) Percussive action (d) Water jet

7. **The underground horizontal way made parallel to strike of the Ore Body is named as-**
   (a) Cross cut (b) Drive
   (c) Adit (d) Shaft

8. **Which underground mining method will be appropriate when host rock and one body both are strong -**
   (a) Sub level stopping (b) Flat backed support
   (c) Shrinkage method (d) Caving method

9. In Rajasthan underground mining method is used at which mine -

(a) Jhamar Kotra mine (b) Palana mine

(c) Giral mine (d) Zawar mine

10. 'Trenching' involves:

(a) Linear excavation

(b) Linear excavation in which one dimension is greater than other

(c) Drilling wedge-shaped boreholes

(d) Digging even-sized pits

11. Diamond drilling can be used to bore holes in:

(a) Horizontal direction only

(b) Vertical direction only

(c) Horizontal and vertical directions

(d) All directions

12. The P-wave velocities are highest in:

(a) Air (b) Water

(c) Sand (d) Granite

13. Which of the following has the least electrical resistivity:

(a) Diamond (b) Sulphur

(c) Clay (d) Anthracite

14. A horizontal entry into an ore body is called:

(a) Adit (b) Shaft

(c) Bench (d) Pit

15. Seismic facies analysis is:

(a) Regional stratigraphic interpretation only

(b) Depositional environment only

(c) Geologic history only

(d) All the three above

16. Seismic attributes represents the seismically driven parameters:

(a) Velocity only (b) Amplitude only

(c) Frequency only (d) All the three above

17. A well log of the travel time (transit time) for the seismic waves is:

(a) Inversely proportional to S-wave velocity
(b) Reciprocal of P-wave velocity
(c) Equal to the S-wave velocity
(d) Equal to the P-wave velocity

18. If the water saturation is S_w, then oil saturation is:

(a) S_w/S.S.P. (b) 1-S_w
(c) S_w^{-1} (d) 1-R_w

19. In seismic method of exploration the recorded waves are of:

(a) Low frequency (b) Zero frequency
(c) High frequency (d) Aliased frequency

20. The radius of the Fresnel zone is given by:

(a) R = dm/2 (b) R = d^2 /m
(c) R=ddm/ (d) R =4rm^2

21. Snell's law is applied to:

(a) Reflected ray only (b) Refracted ray only
(c) Diffracted ray only (d) Reflected and refracted rays both

22. Poisson's ratio for consolidated rocks is about:

(a) 0.25 (b) 0.12
(c) 1.30 (d) 3.01

23. The propagation of P wave and S wave are as under:

(a) P wave arrives later than S wave
(b) P wave and S wave arrive simultaneously
(c) P wave arrives earlier than S wave
(d) S wave arrives earlier than P wave

24. The associated particle motion in Rayleigh wave is:

(a) Circular (b) Elliptic
(c) Straight line (d) Hyperbola

25. Love waves are used for the study of the velocity structure of the:

(a) Lithosphere only
(b) Asthenosphere only

(c) Hydrosphere

(d) Lithosphere and Asthenosphere both

26. Diffraction of seismic waves in the ground is due to:

(a) Edges of faulted layers only

(b) Small isolated objects only

(c) Edges of faulted layers and small objects both

(d) Scattering of incident seismic energy

27. Water holding capacity is more in

(a) Clayey (b) Sandy

(c) Loamy (d) Red soils.

28. The principal of churn drilling is -

(a) Rotary action (b) Rotary action with water jet

(c) Percussive action (d) Water jet

29. The underground horizontal way made parallel to strike of the Ore Body is named as-

(a) Cross cut (b) Drive

(c) Adit (d) Shaft

30. Which type of clay in used as drilling mud?

(a) China Clay (b) Fuller's Earth

(c) Bentonite Clay (d) Fire Clay

31. Which one of the following is not a good reservoir rock for petroleum?

(a) Limestone (b) Sandstone

(c) Mudstone (d) Reef bound stone

32. Weathering of feldspar minerals may result in which one of the following set of product?

(a) Clays, Ca, Na, K ions, dissolved silica

(b) Iron oxides, Ca, Fe, Mg, Mn ions, clays dissolved silica

(c) Fe, Mg ions, Mn oxides, clays

(d) Fe, Mn, Cu and Pb

33. A Dam is an example of which type of system:

(a) Natural (b) Modified

(c) Controlled (d) Uncontrolled

34. The example of an Arch Dam is -

(a) Idduki Dam (b) Jawai Dam

(c) Bhakra Dam (d) Mahi Dam

35. The Sardar Sarovar Dam is a type of -

(a) Arch Dam (b) Gravity Dam

(c) Embankment Dam (d) Coffee Dam

36. The Bilaspur Dam is constructed on which river -

(a) Chambal River (b) Khari River

(c) Banas River (d) Luni River

37. Rutting is a phenomenon associated with:

(a) Bridge (b) Pavements

(c) Dams (d) Tunnels

38. Which one of the following rock types has highest compressive strength?

(a) Diorite (b) Marble

(c) Sandstone (d) Basalt

39. Which one of the following tests is used to know the durability of building stones?

(a) Jack test (b) Smithis test

(c) Los Angeles test (d) Darry test

40. The method which involves the forceful injection of slurry of water and cement into the rock fractured rocks of the site, is known as:

(a) Back filling (b) Lining

(c) Cement stabilization (d) Grouting

41. The walls constructed on both sides of the crest of a dam are known as:

(a) Retaining wall (b) Key wall

(c) Safety wall (d) Parapet wall

42. Which one of the following arrangements made in a dam near the top to let off excess water of the reservoir to the downstream side?

(a) Diversion tunnel (b) Sluice

(c) Cut of wall (d) Spillway

ANSWER SHEET

1. (c)	2. (d)	3. (a)	4. (c)	5. (a)
6. (c)	7. (b)	8. (a)	9. (d)	10. (a)
11. (d)	12. (d)	13. (d)	14. (a)	15. (d)
16. (d)	17. (b)	18. (b)	19. (c)	20. (d)
21. (b)	22. (c)	23. (b)	24. (d)	25. (d)
26. (c)	27. (a)	28. (c)	29. (c)	30. (c)
31. (c)	32. (a)	33. (c)	34. (a)	35. (d)
36. (c)	37. (b)	38. (d)	39. (d)	40. (d)
41. (d)	42. (d)			